I0047958

Noah Porter

Half Hours with Modern Scientists

Noah Porter

Half Hours with Modern Scientists

ISBN/EAN: 9783743345898

Manufactured in Europe, USA, Canada, Australia, Japa

Cover: Foto ©berggeist007 / pixelio.de

Manufactured and distributed by brebook publishing software (www.brebook.com)

Noah Porter

Half Hours with Modern Scientists

HALF HOURS

WITH

MODERN SCIENTISTS.

LECTURES AND ESSAYS

BY

PROFS. HUXLEY, BARKER, STIRLING, COPE AND
TYNDALL.

WITH

A GENERAL INTRODUCTION

BY

NOAH PORTER, D.D., LL.D.,

PRESIDENT OF YALE COLLEGE.

1872.

CONTENTS.

INTRODUCTION TO THE NEW EDITION OF HALF-HOURS WITH MODERN SCIENTISTS.

The title of this Series of Essays—*Half Hours with Modern Scientists*—suggests a variety of thoughts, some of which may not be inappropriate for a brief introduction to a new edition. *Scientist* is a modern appellation which has been specially selected to designate a devotee to one or more branches of physical science. Strictly interpreted it might properly be applied to the student of any department of knowledge when prosecuted in a scientific method, but for convenience it is limited to the student of some branch of physics. It is not thereby conceded that nature, *i. e.*, physical or material nature is any more legitimately or exclusively the field for scientific enquiries than spirit, or that whether the objects of science are material or spiritual, the assumptions and processes of science themselves should not be subjected to scientific analysis and justification. There are so-called philosophers who adopt both these conclusions. There are those who reason and dogmatize as though nature were synonymous with matter, or as though spirit, if there be such an essence, must be conceived and explained after the principles and analogies of matter ;—others assume that a science of scientific method can be nothing better than the mist or moonshine which they vilify by the name of metaphysics. But unfortunately for such opinions the fact is constantly forced upon the attention of scientists of every description, that the agent by which they examine matter is more than matter, and that this agent, whatever be its substance, asserts its prerogatives to determine the conceptions which the scientist forms of matter as well as to the methods by which he investigates material pro-perties. Even the positivist philosopher who not only denounces metaphysics as illegitimate, but also contends that the metaphysical era of human in-quiry, has in the development of scientific progress been outgrown like the measles, which is expe-rienced but once in a life-time; finds when his positivist theory is brought to the test that positiv-ism itself in its very problem and its solutions, is but the last adopted metaphysical theory of science.

We also notice that it is very difficult, if not im-possible, for the inquisitive scientist to limit himself strictly to the object-matter of his own chosen field, and not to enquire more or less earnestly—not in-frequently to dogmatize more or less positively—respecting the results of other sciences and even respecting the foundations and processes of scien-tific inquiry itself. Thus Mr. Huxley in the first Essay of this Series on *The Physical Basis of Life*, leaves the discussion of his appropriate theme in order to deliver sundry very positive and pro-nounced assertions respecting the " limits of philoso-phical inquiry," and quotes with manifest satisfac-tion a dictum of David Hume that is sufficiently dogmatic and positive, as to what these limits are.

In more than one of his Lay sermons, he rushes headlong into the most pronounced assertions in respect to the nature of matter and of spirit. The eloquent Tyndall, in No. 5, expounds at length *The Methods and Tendencies of Physical Investigation* and discourses eloquently, if occasionally somewhat poetically, of *The Scientific use of the Imagination.* But Messrs. Huxley and Tyndall are eminent examples of scientists who are severely and successfully devoted respectively to physiology and the higher physics. No one will contend that they have not faithfully cultivated their appropriate fields of inquiry. The fact that neither can be content to confine himself within his special field, forcibly illustrates the tendency of every modern science to concern itself with its relations to its neighbors, and the unresistible necessity which forces the most rigid physicist to become a metaphysician in spite of himself. So much for the appellation " *Scientists.*"

"*Half Hours*" suggests the very natural inquiry —What can a scientist communicate in half an hour, especially to a reader who may be ignorant of the elements of the science which he would expound? Does not the phrase *Half Hours with Modern Scientists* stultify itself and suggest the folly of any attempt to treat of science with effect in a series of essays? In reply we would ask the attention of the reader to the following considerations.

The tendency is universal among the scientific men of all nations, to present the principles of science in such brief summaries or statements as may bring them within the reach of common readers.

The tendency indicates that there is a large body of readers who are so far instructed in the elements of science as to be able to understand these summaries. In England, Germany, France and this country such brief essays are abundant, either in the form of contributions to popular and scientific journals, or in that of popular lectures, or in that of brief manuals, or of monographs on separate topics; especially such topics as are novel, or are interesting to the public for their theoretic brilliancy, or their applications to industry and art.

These essays need not be and they are not always superficial, because they are brief. They often are the more profound on account of their conciseness, as when they contain a condensed summary of the main principles of the art or science in question, or a brief history of the successive experiments which have issued in some brilliant discovery. These essays are very generally read, even though they are both concise and profound. But they could not be read even though they were less profound than they are, were there not provided a numerous company of readers who are sufficiently instructed in science to appreciate them. That such a body of readers exists in the countries referred to, is easily explained by the existence of public schools and schools of science and technology, by the enormous extension of the knowledge of machinery, engineering, mining, dyeing, etc., etc., all of which imply a more or less distinct recognition of scientific principles and stimulate the curiosity in regard to scientific truth. Popular lectures also, illustrated by experiments, have been repeated before thousands of excited listeners, and the eager and inventive minds of multitudes of ingenious youths have been trained by this distribution of science, to the capacity to comprehend the compact and pointed scientific essay, even though it taxes the attention and suspends the breath for a half-hour by its closeness and severity.

The fact is also worthy of notice, that many of the ablest scientists of our times have made a special study of the art of expounding and presenting scientific truth. Some of them have schooled themselves to that lucid and orderly method by which a science seems to spring into being a second time, under the creative hand of its skilful expositor. Others have made a special study of philosophic diction. Others have learned how to adorn scientific truth with the embellishments of an affluent imagination. Some of the ablest writers of our time are found among the devotees of physical science. That a few scientific writers and lecturers may have exemplified some of the most offensive features of the demagogue and the sophist cannot be denied, but we may not forget that many have attained to the consummate skill of the accomplished essayist and impressive and eloquent orator.

One advantage cannot be denied of this now popular and established method of setting forth scientific truth, viz., that it prescribes a convenient method of bringing into contrast the arguments *for* and *against* any disputed position in science. If materialism can furnish its ready advocate with a convenient vehicle for its ready diffusion, the antagonist theory can avail itself of a similar vehicle for the communication of the decisive and pungent reply. The one is certain to call forth the other, and if the two are present side by side in the same series, so much the better is it for the truth and so much the worse for the error. The teacher before his class, the lecturer in the presence of his audience, has the argument usually to himself; he allows few questionings and admits no reply. An erroneous theory may entrench itself within a folio against arguments which would annihilate its positions if these were condensed in a tract.

This consideration should dispel all the alarm that is felt by the defenders of religion in view of the general diffusion of popular scientific treatises. The brief statement of a false or groundless scientific theory, even by its defender, is often its most effectual refutation. A magnificently imposing argument often shrinks into insignificance when its advocate is forced to state its substance in a compact and close-jointed outline. The articulations are seen to be defective, the joints do not fit one another, the coherence is conspicuously wanting. Let then error do its utmost in the field of science. Its deficient data and its illogical processes are certain to be exposed, sometimes even by its own advocates. If this does not happen the defender of that scientific truth which seems to be essential to the teachings and faiths of religion, must scrutinize its reasonings by the rules and methods of scientific inquiry. If science seems to be hostile to religion, this very seeming should arouse the defender of Theism and Christianity to examine into the grounds both by the light and methods which are appropriate to science itself. The more brief and compact and popular is the argument which he is to refute, the more feasible is the task of exposure and reply. Only let this be a cardinal maxim with the defender of the truth, that whatever is scientifically defended and maintained must be scientifically refuted and overthrown. The great Master of our faith never uttered a more comprehensive or a grander maxim than the memorable words, " *To this end was I born and for this cause came I into the world, that I should bear witness unto the truth. Everyone that is of the truth heareth my voice.*" It would be easy to show that the belief in moral and religious truth and the freedom in searching for and defending it which was inspired by these words have been most efficient in training the human mind to that faith in the results of scientific investigation which characterize the modern scientist. That Christian believer must either have a very imperfect view of the spirit of his own faith, or a very narrow conception of the evidences and the effect of its teachings, who imagines that the freest spirit of scientific inquiry, or the most penetrating insight into the secrets of matter or of spirit can have any other consequence than to strengthen and brighten the evidence for Christian truth.

N. P.

Yale College, *May*, 1872,

PUBLISHERS' NOTE TO SECOND EDITION.

The five lectures embodied in this First Series of Half Hours with Modern Scientists were first published as Nos. I.—V. of the University Scientific Series. In this series the publishers have aimed to give to the public in a cheap pamphlet form, the advance thought in the Scientific world. The intrinsic value of these lectures has created a very general desire to have them put in a permanent form. They therefore have brought them out in this style. Each five succeeding numbers of this celebrated series will be printed and bound in uniform style with this volume, and be designated as second series, third series, and so on. Henceforth it will be the design of the publishers to give preference to those lectures and essays of American scientists which contain original research and discovery, rather than to reprinting from European sources. The lectures in the second series will be (1) On Natural Selection as Applied to Man, by Alfred Russel Wallace; (2) three profoundly interesting lectures on Spectrum Analysis, by Profs. Roscoe, Huggins, and Lockyer; (3) the Sun and its Different Atmospheres, a lecture by Prof. C. A. Young, Ph.D., of Dartmouth College; (4) the Earth a great Magnet, by Prof. A. M. Mayer, Ph.D., of Stevens Institute; and (5) the Mysteries of the Voice and Ear, by Prof. Ogden N. Rood, of Columbia College. The last three lectures contain many original discoveries and brilliant experiments, and are finely illustrated.

ON THE PHYSICAL BASIS OF LIFE.

INTRODUCTION.

The following remarkable discourse was originally delivered in Edinburg, November 18th, 1868, as the first of a series of Sunday evening addresses, upon non-religious topics, instituted by the Rev. J. Cranbrook. It was subsequently published in London as the leading article in the *Fortnightly Review*, for February, 1869, and attracted so much attention that five editions of that number of the magazine have already been issued. It is now re-printed in this country, in permanent form, for the first time, and will doubtless prove of great interest to American readers. The author is Thomas Henry Huxley, of London, Professor of Natural History in the Royal School of Mines, and of Comparative Anatomy and Physiology in the Royal College of Surgeons. He is also President of the Geological Society of London. Although comparatively a young man, his numerous and valuable contributions to Natural Science entitle him to be considered one of the first of living Naturalists, especially in the departments of Zoölogy and Paleontology, to which he has mainly devoted himself. He is undoubtedly the ablest English advocate of Darwin's theory of the Origin of Species, particularly with reference to its application to the human race, which he believes to be nearly related to the higher apes. It is, indeed, through his discussion of this question that he is, perhaps, best known to the general public, as his late work entitled "Man's Place in Nature," and other writings on similar topics, have been very widely read in this country and in Europe. In the present lecture Professor Huxley discusses a kindred subject of no less interest and importance, and should have an equally candid hearing.

YALE COLLEGE, *March 30th,* 1869.

On the Physical Basis of Life.

In order to make the title of this discourse generally intelligible, I have translated the term "Protoplasm," which is the scientific name of the substance of which I am about to speak, by the words "the physical basis of life." I suppose that, to many, the idea that there is such a thing as a physical basis, or matter, of life may be novel—so widely spread is the conception of life as a something which works through matter, but is independent of it; and even those who are aware that matter and life are inseparably connected, may not be prepared for the conclusion plainly suggested by the phrase "the physical basis or matter of life," that there is some one kind of matter which is common to all living beings, and that their endless diversities are bound together by a physical, as well as an ideal, unity. In fact, when first apprehended, such a doctrine as this appears almost shocking to common sense. What, truly, can seem to be more obviously different from one another in faculty, in form, and in substance, than the various kinds of living beings? What community of faculty can there be between the brightly-colored lichen, which so nearly resembles a mere mineral incrustation of the bare rock on which it grows, and the painter, to whom it is instinct with beauty, or the botanist, whom it feeds with knowledge?

Again, think of the microscopic fungus—a mere infinitesimal ovoid particle, which finds space and duration enough to multiply into countless millions in the body of a living fly; and then of the wealth of foliage, the luxuriance of flower and fruit, which lies between this bald sketch of a plant and the giant pine of California, towering to the dimensions of a cathedral spire, or the Indian fig, which covers acres with its profound shadow, and endures while nations and empires come and go around its vast circumference! Or, turning to the other half of the world of life, picture to yourselves the great finner whale, hugest of beasts that live, or have lived, disporting his eighty or ninety feet of bone, muscle and blubber, with easy roll, among waves in which the stoutest ship that ever left dockyard would founder hopelessly; and contrast him with the invisible animalcules—mere gelatinous specks, multitudes of which could, in fact, dance upon the point of a needle with the same ease as the angels of the schoolmen could, in imagination. With these images before your minds, you may well ask what community of form, or structure, is there between the animalcule and the whale, or between the fungus and fig-tree? And, *a fortiori*, between all four?

Finally, if we regard substance, or material composition, what hidden bond can connect the flower which a girl wears in her hair and the blood which courses through her youthful veins; or, what is there in common between the dense and resisting mass of the oak, or the strong fabric of the tortoise, and those broad disks of glassy

jelly which may be seen pulsating through the waters of
a calm sea, but which drain away to mere films in the
hand which raises them out of their element? Such objections as these must, I think, arise in the mind of every
one who ponders, for the first time, upon the conception
of a single physical basis of life underlying all the diversities of vital existence ; but I propose to demonstrate
to you that, notwithstanding these apparent difficulties,
a threefold unity—namely, a unity of power or faculty,
a unity of form, and a unity of substantial composition—
does pervade the whole living world. No very abstruse
argumentation is needed, in the first place, to prove that
the powers, or faculties, of all kinds of living matter, diverse as they may be in degree, are substantially similar
in kind. Goethe has condensed a survey of all the powers of mankind into the well-known epigram :

" Warum treibt sich das Volk so und schreit ? Es will sich ernähren
    Kinder zeugen, und sie nähren so gut es vermag.
*   *   *   *   *   *   *   *   *
    Weiter bringt es kein Mensch, stell' er sich, wie er auch will."

In physiological language this means, that all the multifarious and complicated activities of man are compre-
hensible under three categories. Either they are immediately directed towards the maintenance and development of the body, or they effect transitory changes
in the relative positions of parts of the body, or they
tend towards the continuance of the species. Even
those manifestations of intellect, of feeling, and of will,
which we rightly name the higher faculties, are not excluded from this classification, inasmuch as to every one
but the subject of them, they are known only as transit-

ory changes in the relative positions of parts of the body. Speech, gesture, and every other form of human action are, in the long run, resolvable into muscular contraction, and muscular contraction is but a transitory change in the relative positions of the parts of a muscle. But the scheme, which is large enough to embrace the activities of the highest form of life, covers all those of the lower creatures. The lowest plant, or animalcule, feeds, grows and reproduces its kind. In addition, all animals manifest those transitory changes of form which we class under irritability and contractility; and it is more than probable, that when the vegetable world is thoroughly explored, we shall find all plants in possession of the same powers, at one time or other of their existence. I am not now alluding to such phenomena, at once rare and conspicuous, as those exhibited by the leaflets of the sensitive plant, or the stamens of the barberry, but to much more widely-spread, and, at the same time, more subtle and hidden, manifestations of vegetable contractility. You are doubtless aware that the common nettle owes its stinging property to the innumerable stiff and needle-like, though exquisitely delicate, hairs which cover its surface. Each stinging-needle tapers from a broad base to a slender summit, which, though rounded at the end, is of such microscopic fineness that it readily penetrates, and breaks off in, the skin. The whole hair consists of a very delicate outer case of wood, closely applied to the inner surface of which is a layer of semi-fluid matter, full of innumerable granules of extreme minuteness. This semi-fluid lining is protoplasm, which thus constitutes a kind of bag, full of a limpid liquid, and roughly corresponding in form with the interior of the hair which it fills. When viewed with a sufficiently high magnifying power, the protoplasmic layer of the nettle hair is seen to be in a condition of unceasing activity. Local contractions of the whole thickness of its substance pass slowly and gradually from point to point, and give rise to the appearance of progressive waves, just as the bending of successive stalks of corn by a breeze produces the apparent billows of a corn-field. But, in addition to these movements, and independently of them, the granules are driven, in relatively rapid streams, through channels in the protoplasm which seem to have a considerable amount of persistence. Most commonly, the currents in adjacent parts of the protoplasm take similar directions; and, thus, there is a general stream up one side of the hair and down the other. But this does not prevent the existence of partial currents which take different routes; and, sometimes, trains of granules may be seen coursing swiftly in opposite directions, within a twenty-thousandth of an inch of one another; while, occasionally, opposite streams come into direct collision, and, after a longer or shorter struggle, one predominates. The cause of these currents seem to lie in contractions of the protoplasm which bounds the channels in which they flow, but which are so minute that the best microscopes show only their effects, and not themselves.

The spectacle afforded by the wonderful energies prisoned within the compass of the microscopic hair of a plant, which we commonly regard as a merely passive organism, is not easily forgotten by one who has watched its display continued hour after hour, without pause or sign of weakening. The possible complexity of many other organic forms, seemingly as simple as the protoplasm of the nettle, dawns upon one; and the comparison of such a protoplasm to a body with an internal circulation, which has been put forward by an eminent physiologist, loses much of its startling character. Currents similar to those of the hairs of the nettle have been observed in a great multitude of very different plants, and weighty authorities have suggested that they probably occur, in more or less perfection, in all young vegetable cells. If such be the case, the wonderful noonday silence of a tropical forest is, after all, due only to the dullness of our hearing; and could our ears catch the murmur of these tiny maelstroms, as they whirl in the innumerable myriads of living cells which constitute each tree, we should be stunned, as with the roar of a great city.

Among the lower plants, it is the rule rather than the exception, that contractility should be still more openly manifested at some periods of their existence. The protoplasm of *Algæ* and *Fungi* becomes, under many circumstances, partially, or completely, freed from its woody case, and exhibits movements of its whole mass, or is propelled by the contractility of one or more hairlike prolongations of its body, which are called vibratile cilia. And, so far as the conditions of the manifestation of the phenomena of contractility have yet been studied, they are the same for the plant as for the animal. Heat and electric shocks influence both, and in the same way, though it may be in different degrees. It is by no means my intention to suggest that there is no difference in faculty between the lowest plant and the highest, or between plants and animals. But the difference between the powers of the lowest plant, or animal, and those of the highest is one of degree, not of kind, and depends, as Milne-Edwards long ago so well pointed out, upon the extent to which the principle of the division of labor is carried out in the living economy. In the lowest organism all parts are competent to perform all functions, and one and the same portion of protoplasm may successively take on the function of feeding, moving, or reproducing apparatus. In the highest, on the contrary, a great number of parts combine to perform each function, each part doing its allotted share of the work with great accuracy and efficiency, but being useless for any other purpose. On the other hand, notwithstanding all the fundamental resemblances which exist between the powers of the protoplasm in plants and in animals, they present a striking difference (to which I shall advert more at length presently,) in the fact that plants can manufacture fresh protoplasm out of mineral compounds, whereas animals are obliged to procure it ready made, and hence, in the long run, depend upon plants. Upon what condition this difference in the powers of the two great divisions of the world of life depends, nothing is at present known.

With such qualification as arises out of the last-mentioned fact, it may be truly said that the acts of all living things are fundamentally one. Is any such unity predicable of their forms? Let us seek in easily verified facts for a reply to this question. If a drop of blood be drawn by pricking one's finger, and viewed with proper precautions and under a sufficiently high microscopic power, there will be seen, among the innumerable multitude of little, circular, discoidal bodies, or corpuscles, which float in it and give it its color, a comparatively small number of colorless corpuscles, of somewhat larger size and very irregular shape. If the drop of blood be kept at the temperature of the body, these colorless corpuscles will be seen to exhibit a marvelous activity, changing their forms with great rapidity, drawing in and thrusting out prolongations of their substance, and creeping about as if they were independent organisms. The substance which is thus active is a mass of protoplasm, and its activity differs in detail, rather than in principle, from that of the protoplasm of the nettle. Under sundry circumstances the corpuscle dies and becomes distended into a round mass, in the midst of which is seen a smaller spherical body, which existed, but was more or less hidden, in the living corpuscle, and is called its *nucleus*. Corpuscles of essentially similar structure are to be found in the skin, in the lining of the mouth, and scattered through the whole frame work of the body. Nay, more; in the earliest condition of the human organism, in that state in which it has just become distinguishable from the egg in which it arises, it is nothing but an aggregation of such corpuscles, and every organ of the body was, once, no more than such an aggregation. Thus a nucleated mass of protoplasm turns out to be what may be termed the structural unit of the human body. As a matter of fact, the body, in its earliest state, is a mere multiple of such units; and, in its perfect condition, it is a multiple of such units, variously modified. But does the formula which expresses the essential structural character of the highest animal cover all the rest, as the statement of its powers and faculties covered that of all others? Very nearly. Beast and fowl, reptile and fish, mollusk, worm, and polype, are all composed of structural units of the same character, namely, masses of protoplasm with a nucleus. There are sundry very low animals, each of which, structurally, is a mere colorless blood-corpuscle, leading an independent life. But, at the very bottom of the animal scale, even this simplicity becomes simplified, and all the phenomena of life are manifested by a particle of protoplasm without a nucleus. Nor are such organisms insignificant by reason of their want of complexity. It is a fair question whether the protoplasm of those simplest forms of life, which people an immense extent of the bottom of the sea, would not outweigh that of all the higher living beings which inhabit the land, put together. And in ancient times, no less than at the present day, such living beings as these have been the greatest of rock builders.

What has been said of the animal world is no less true of plants. Imbedded in the protoplasm at the broad, or attached, end of the nettle hair, there lies a spheroidal nucleus. Careful examination further proves that the whole substance of the nettle is made up of a repetition of such masses of nucleated protoplasm, each contained in a wooden case, which is modified in form, sometimes into a woody fibre, sometimes into a duct or spiral vessel, sometimes into a pollen grain, or an ovule. Traced back to its earliest state, the nettle arises as the man does, in a particle of nucleated protoplasm. And in the lowest plants, as in the lowest animals, a single mass of such protoplasm may constitute the whole plant, or the protoplasm may exist without a nucleus. Under these circumstances it may well be asked, how is one mass of non-nucleated protoplasm to be distinguished from another? why call one "plant" and the other "animal?" The only reply is that, so far as form is concerned, plants and animals are not separable, and that, in many cases, it is a mere matter of convention whether we call a given organism an animal or a plant.

There is a living body called *Æthalium septicum*, which appears upon decaying vegetable substances, and in one of its forms, is common upon the surface of tan pits. In this condition it is, to all intents and purposes, a fungus, and formerly was always regarded as such; but the remarkable investigations of De Bary have shown that, in another condition, the *Æthalium* is an actively locomotive creature, and takes in solid matters, upon which, apparently, it feeds, thus exhibiting the most characteristic feature of animality. Is this a plant, or is it an animal? Is it both, or is it neither? Some decide in favor of the last supposition, and establish an intermediate kingdom, a sort of biological No Man's Land for all these questionable forms. But, as it is admittedly impossible to draw any distinct boundary line between this no man's land and the vegetable world on the one hand, or the animal, on the other, it appears to me that this proceeding merely doubles the difficulty which, before, was single. Protoplasm, simple or nucleated, is the formal basis of all life. It is the clay of the potter;

which, bake it and paint it as he will, remains clay, separated by artifice, and not by nature, from the commonest brick or sun-dried clod. Thus it becomes clear that all living powers are cognate, and that all living torms are fundamentally of one character.

The researches of the chemist have revealed a no less striking uniformity of material composition in living matter. In perfect strictness, it is true that chemical investigation can tell us little or nothing, directly, of the composition of living matter, inasmuch as such matter must needs die in the act of analysis, and upon this very obvious ground, objections, which I confess seem to me to be somewhat frivolous, have been raised to the drawing of any conclusions whatever respecting the composition of actually living matter from that of the dead matter of life, which alone is accessible to us. But objectors of this class do not seem to reflect that it is also, in strictness, true that we know nothing about the composition of any body whatever, as it is. The statement that a crystal of calc-spar consists of carbonate of lime, is quite true, if we only mean that, by appropriate processes, it may be resolved into carbonic acid and quicklime. If you pass the same carbonic acid over the very quicklime thus obtained, you will obtain carbonate of lime again ; but it will not be calc-spar, nor anything like it. Can it, therefore, be said that chemical analysis teaches nothing about the chemical composition of calc-spar ? Such a statement would be absurd ; but it is hardly more so than the talk one occasionally hears about the uselessness of applying the results of chemical analysis to the living bodies which have yielded them. One fact, at

any rate, is out of reach of such refinements, and this
is, that all the forms of protoplasm which have yet been
examined contain the four elements, carbon, hydrogen,
oxygen, and nitrogen, in very complex union, and that
they behave similarly towards several re-agents. To this
complex combination, the nature of which has never
been determined with exactness, the name of Protein
has been applied. And if we use this term with such
caution as may properly arise out of our comparative
ignorance of the things for which it stands, it may be
truly said, that all protoplasm is proteinaceous ; or, as
the white, or albumen, of an egg is one of the commonest examples of a nearly pure proteine matter, we may
say that all living matter is more or less albuminoid.
Perhaps it would not yet be safe to say that all forms of
protoplasm are affected by the direct action of electric
shocks ; and yet the number of cases in which the contraction of protoplasm is shown to be affected by this
agency increases, every day. Nor can it be affirmed with
perfect confidence that all forms of protoplasm are liable
to undergo that peculiar coagulation at the temperature
of 40 degrees—50 degrees centigrade, which has been
called "heat-stiffening," though Kühne's beautiful researches have proved this occurrence to take place in so
many and such diverse living beings, that it is hardly rash
to expect that the law holds good for all. Enough has,
perhaps, been said to prove the existence of a general
uniformity in the character of the protoplasm, or physical basis of life, in whatever group of living beings it
may be studied. But it will be understood that this general uniformity by no means excludes any amount of

special modifications of the fundamental substance. The mineral, carbonate of lime, assumes an immense diversity of characters, though no one doubts that under all these Protean changes it is one and the same thing. And now, what is the ultimate fate, and what the origin of the matter of life? Is it, as some of the older naturalists supposed, diffused throughout the universe in molecules, which are indestructible and unchangeable in themselves; but, in endless transmigration, unite in innumerable permutations, into the diversified forms of life we know? Or, is the matter of life composed of ordinary matter, differing from it only in the manner in which its atoms are aggregated? Is it built up of ordinary matter, and again resolved into ordinary matter when its work is done? Modern science does not hesitate a moment between these alternatives. Physiology writes over the portals of life,

"Debemur morti nos nostraque,"

with a profounder meaning than the Roman poet attached to that melancholy line. Under whatever disguise it takes refuge, whether fungus or oak, worm or man, the living protoplasm not only ultimately dies and is resolved into its mineral and lifeless constituents, but is always dying, and, strange as the paradox may sound, could not live unless it died. In the wonderful story of the " Peau de Chagrin," the hero becomes possessed of a magical wild ass's skin, which yields him the means of gratifying all his wishes. But its surface represents the duration of the proprietor's life; and for every satisfied desire the skin shrinks in proportion to the intensity of fruition, until at length life and the last handbreadth of the

" Peau de Chagrin," disappear with the gratification of a last wish. Balzac's studies had led him over a wide range of thought and speculation, and his shadowing forth of physiological truth in this strange story may have been intentional. At any rate, the matter of life is a veritable " Peau de Chagrin," and for every vital act it is somewhat the smaller. All work implies waste, and the work of life results, directly or indirectly, in the waste of protoplasm. Every word uttered by a speaker costs him some physical loss ; and, in the strictest sense, he burns that others may have light — so much eloquence, so much of his body resolved into carbonic acid, water and urea. It is clear that this process of expenditure cannot go on forever. But, happily, the protoplasmic *peau de chagrin* differs from Balzac's in its capacity of being repaired, and brought back to its full size, after every exertion. For example, this present lecture, whatever its intellectual worth to you, has a certain physical value to me, which is, conceivably, expressible by the number of grains of protoplasm and other bodily substance wasted in maintaining my vital processes during its delivery. My *peau de chagrin* will be distinctly smaller at the end of the discourse than it was at the beginning. By-and-by, I shall probably have recourse to the substance commonly called mutton, for the purpose of stretching it back to its original size. Now this mutton was once the living protoplasm, more or less modified, of another animal—a sheep. As I shall eat it, it is the same matter altered, not only by death, but by exposure to sundry artificial operations in the process of cooking. But these changes, whatever be their extent,

have not rendered it incompetent to resume its old functions as matter of life. A singular inward laboratory,
which I possess, will dissolve a certain portion of the
modified protoplasm, the solution so formed will pass
into my veins ; and the subtle influences to which it will
then be subjected wifl convert the dead protoplasm into
living protoplasm, and transubstantiate sheep into man.
Nor is this all. If digestion were a thing to be trifled
with, I might sup upon lobster, and the matter of life of
the crustacean would undergo the same wonderful metamorphosis into humanity. And were I to return to my
own place by sea, and undergo shipwreck, the crustacea
might, and probably would, return the compliment, and
demonstrate our common nature by turning my protoplasm into living lobster. Or, if nothing better were to
be had, I might supply my wants with mere bread, and I
should find the protoplasm of the wheat-plant to be convertible into man, with no more trouble than that of the
sheep, and with far less, I fancy, than that of the lobster.
Hence it appears to be a matter of no great moment what
animal, or what plant, I lay under contribution for protoplasm, and the fact speaks volumes for the general identity of that substance in all living beings. I share this
catholicity of assimilation with other animals, all of
which, so far as we know, could thrive equally well on the
protoplasm of any of their fellows, or of any plant ; but
here the assimilative powers of the animal world cease.

A solution of smelling-salts in water with an infinitesimal proportion of some other saline matters, contains
all the elementary bodies which enter into the composition of protoplasm ; but, as I need hardly say, a hogs-

head of that fluid would not keep a hungry man from starving, nor would it save any animal whatever from a like fate. An animal cannot make protoplasm, but must take it ready-made from some other animal, or some plant —the animal's highest feat of constructive chemistry being to convert dead protoplasm into that living matter of life which is appropriate to itself. Therefore, in seeking for the origin of protoplasm, we must eventually turn to the vegetable world. The fluid containing carbonic acid, water, and ammonia, which offers such a barmecide feast to the animal, is a table richly spread to multitudes of plants; and with a due supply of only such materials, many a plant will not only maintain itself in vigor, but grow and multiply until it has increased a million-fold, or a million million-fold, the quantity of protoplasm which it originally possessed; in this way building up the matter of life, to an indefinite extent, from the common matter of the universe. Thus the animal can only raise the complex substance of dead protoplasm to the higher power, as one may say, of living protoplasm; while the plant can raise the less complex substances— carbonic acid, water, and ammonia—to the same stage of living protoplasm, if not to the same level. But the plant also has its limitations. Some of the fungi, for example, appear to need higher compounds to start with, and no known plant can live upon the uncompounded elements of protoplasm. A plant supplied with pure carbon, hydrogen, oxygen, and nitrogen, phosphorus, sulphur, and the like, would as infallibly die as the animal in his bath of smelling-salts, though it would be surrounded by all the constituents of protoplasm. Nor, indeed, need the process of simplification of vegetable food be carried so far as this, in order to arrive at the limit of the plant's thaumaturgy.

Let water, carbonic acid, and all the other needful constituents, be supplied without ammonia, and an ordinary plant will still be unable to manufacture protoplasm. Thus the matter of life, so far as we know it (and we have no right to speculate on any other) breaks up in consequence of that continual death which is the condition of its manifesting vitality, into carbonic acid, water, and ammonia, which certainly possess no properties but those of ordinary matter; and out of these same forms of ordinary matter and from none which are simpler, the vegetable world builds up all the protoplasm which keeps the animal world agoing. Plants are the accumulators of the power which animals distribute and disperse.

But it will be observed, that the existence of the matter of life depends on the preëxistence of certain compounds, namely, carbonic acid, water, and ammonia. Withdraw any one of these three from the world and all vital phenomena come to an end. They are related to the protoplasm of the plant, as the protoplasm of the plant is to that of the animal. Carbon, hydrogen, oxygen, and nitrogen are all lifeless bodies. Of these, carbon and oxygen unite in certain proportions and under certain conditions, to give rise to carbonic acid; hydrogen and oxygen produce water; nitrogen and hydrogen give rise to ammonia. These new compounds, like the elementary bodies of which they are composed, are lifeless. But when they are brought together, under certain conditions they give rise to the still more complex body, protoplasm, and this protoplasm exhibits the phenomena of life. I see no break in this series of steps in molecular complication, and I am unable to understand why the language which is applicable to any one term of the series may not be used to any of the others. We think fit to call different kinds of matter carbon, oxygen, hydrogen, and nitrogen, and to speak of the various powers and activities of these substances as the properties of the matter of which they are composed. When hydrogen and oxygen are mixed in a certain proportion, and the electric spark is passed through them, they disappear and a quantity of water, equal in weight to the sum of their weights, appears in their place. There is not the slightest parity between the passive and active powers of the water and those of the oxygen and hydrogen which have given rise to it. At 32 degrees Fahrenheit, and far below that temperature, oxygen and hydrogen are elastic gaseous bodies, whose particles tend to rush away from one another with great force. Water, at the same temperature, is a strong though brittle solid, whose particles tend to cohere into definite geometrical shapes, and sometimes build up frosty imitations of the most complex forms of vegetable foliage. Nevertheless we call these, and many other strange phenomena, the properties of the water, and we do not hesitate to believe that, in some way or another, they result from the properties of the component elements of the water. We do not assume that a something called "aquosity" entered into and took possession of the oxide of hydrogen as soon as it was formed, and then guided the aqueous particles to their places in the facets of the crystal, or amongst the leaflets of the hoar-frost. On the contrary, we live in the hope and in the faith that, by the advance of molecular physics, we shall by-and-by be able to see our way as clearly from the constituents of water to the properties of water, as we are now able to deduce the operations of a watch from the form of its parts and the manner in which they are put together. Is the case in any way changed when carbonic acid, water and ammonia disappear, and in their place, under the influence of preëxisting living protoplasm, an equivalent weight of the matter of life makes its appearance? It is true that there is no sort of parity between the properties of the components and the properties of the resultant, but neither was there in the case of the water. It is also true that what I have spoken of as the influence of preëxisting living matter is something quite unintelligible ; but does any body quite comprehend the *modus operandi* of an electric spark, which traverses a mixture of oxygen and hydrogen? What justification is there, then, for the assumption of the existence in the living matter of a something which has no representative or correlative in the not living matter which gave rise to it? What better philosophical status has " vitality " than " aquosity?" And why should "vitality" hope for a better fate than the other "itys" which have disappeared since Martinus Scriblerus accounted for the operation of the meat-jack by its inherent " meat roasting quality," and scorned the "materialism " of those who explained the turning of the spit by a certain mechanism worked by the draught of the chimney? If scientific language is to possess a definite and constant signification whenever it is employed, it seems to me that we are logically bound to apply to the protoplasm, or physical basis of life, the same conceptions as those which are held to be legitimate elsewhere. If the phenomena exhibited by water are its properties, so are those presented by protoplasm, living or dead, its properties. If the properties of water may be properly said to result from the nature and disposition of its component molecules, I can find no intelligible ground for refusing to say that the properties of protoplasm result from the nature and disposition of its molecules. But I bid you beware that, in accepting these conclusions, you are placing your feet on the first rung of a ladder which, in most people's estimation, is the reverse of Jacob's, and leads to the antipodes of heaven. It may seem a small thing to admit that the dull vital actions of a fungus, or a foraminifer, are the properties of their protoplasm, and are the direct results of the nature of the matter of which they are composed.

But if, as I have endeavored to prove to you, their protoplasm is essentially identical with, and most readily converted into, that of any animal, I can discover no logical halting place between the admission that such is the case, and the further concession that all vital action may, with equal propriety, be said to be the result of the molecular forces of the protoplasm which displays it. And if so, it must be true, in the same sense and to the same extent, that the thoughts to which I am now giving utterance, and your thoughts regarding them, are the expression of molecular changes in that matter of life which is the source of our other vital phenomena. Past experience leads me to be tolerably certain that, when the propositions I have just placed before you are accessible to public comment and criticism, they will be condemned by many zealous persons, and perhaps by some few of the wise and thoughtful. I should not wonder if "gross and brutal materialism" were the mildest phrase applied to them in certain quarters. And most undoubtedly the terms of the propositions are distinctly materialistic. Nevertheless, two things are certain : the one, that I hold the statements to be substantially true ; the other, that I, individually, am no materialist, but, on the contrary, believe materialism to involve grave philosophical error.

This union of materialistic terminology with the repudiation of materialistic philosophy I share with some of the most thoughtful men with whom I am acquainted. And, when I first undertook to deliver the present discourse, it appeared to me to be a fitting opportunity to explain how such an union is not only consistent with, but necessitated by sound logic. I purposed to lead you through the territory of vital phenomena to the materialistic slough in which you find yourselves now plunged, and then to point out to you the sole path by which, in my judgment, extrication is possible. An occurrence, of which I was unaware until my arrival here last night, renders this line of argument singularly opportune. I found in your papers the eloquent address " On the Limits of Philosophical Inquiry," which a distinguished prelate of the English Church delivered before the members of the Philosophical Institution on the previous day. My argument, also, turns upon this very point of

limits of philosophical inquiry; and I cannot bring out
my own views better than by contrasting them with
those so plainly, and, in the main, fairly stated by the
Archbishop of York. But I may be permitted to make
a preliminary comment upon an occurrence that greatly
astonished me. Applying the name of "the New Philosophy" to that estimate of the limits of philosophical
inquiry which I, in common with many other men of science, hold to be just, the Archbishop opens his address
by identifying this "new philosophy" with the positive
philosophy of M. Comte (of whom he speaks as its "founder"); and then proceeds to attack that philosopher and
his doctrine vigorously. Now, so far as I am concerned,
the most Reverend prelate might dialectically hew M.
Comte in pieces, as a modern Agag, and I should not
attempt to stay his hand. In so far as my study of what
specially characterizes the Positive Philosophy has led
me, I find therein little or nothing of any scientific value,
and a great deal which is as thoroughly antagonistic to
the very essence of science as anything in ultramontane Catholicism. In fact, M. Comte's philosophy in
practice might be compendiously described as Catholicism *minus* Christianity. But what has Comptism to do
with the "New Philosophy," as the Archbishop defines
it in the following passage?

"Let me briefly remind you of the leading principles
of this new philosophy.

"All knowledge is experience of facts acquired by the
senses. The traditions of older philosophies have ob-
scured our experience by mixing with it much that the
senses cannot observe, and until these additions are dis-

carded our knowledge is impure. Thus, metaphysics tells us that one fact which we observe is a cause, and another is the effect of that cause ; but upon a rigid analysis we find that our senses observe nothing of cause or effect ; they observe, first, that one fact succeeds another, and, after some opportunity, that this fact has never failed to follow—that for cause and effect we should substitute invariable succession. An older philosophy teaches us to define an object by distinguishing its essential from its accidental qualities ; but experience knows nothing of essential and accidental ; she sees only that certain marks attach to an object, and, after many observations, that some of them attach invariably, whilst others may at times be absent. * * * * * As all knowledge is relative, the notion of anything being necessary must be banished with other traditions."

There is much here that expresses the spirit of the " New Philosophy," if by that term be meant the spirit of modern science ; but I cannot but marvel that the assembled wisdom and learning of Edinburg should have uttered no sign of dissent, when Comte was declared to be the founder of these doctrines. No one will accuse Scotchmen of habitually forgetting their great countrymen ; but it was enough to make David Hume turn in his grave, that here, almost within ear-shot of his house, an instructed audience should have listened, without a murmur, while his most characteristic doctrines were attributed to a French writer of fifty years later date, in whose dreary and verbose pages we miss alike the vigor of thought and the exquisite clearness of the style of the man whom I make bold to term the most acute thinker of the eighteenth century—even though that century produced Kant. But I did not come to Scotland to vindicate the honor of one of the greatest men she has ever produced. My business is to point out to you that the only way of escape out of the crass materialism in which we just now landed is the adoption and strict working out of the very principles which the Archbishop holds up to reprobation.

Let us suppose that knowledge is absolute, and not relative, and therefore, that our conception of matter represents that which it really is. Let us suppose, further, that we do know more of cause and effect than a certain definite order of succession among facts, and that we have a knowledge of the necessity of that succession— and hence, of necessary laws—and I, for my part, do not see what escape there is from utter materialism and necessitarianism. For it is obvious that our knowledge of what we call the material world is, to begin with, at least as certain and definite as that of the spiritual world, and that our acquaintance with the law is of as old a date as our knowledge of spontaneity.

Further, I take it to be demonstrable that it is utterly impossible to prove that anything whatever may not be the effect of a material and necessary cause, and that human logic is equally incompetent to prove that any act is really spontaneous. A really spontaneous act is one which, by the assumption, has no cause ; and the attempt to prove such a negative as this is, on the face of the matter, absurd. And while it is thus a philosophical impossibility to demonstrate that any given phenomenon is not the effect of a material cause, any one who is acquainted with the history of science will admit, that its progress has, in all ages, meant, and now more than ever means, the extension of the province of what we call matter and causation, and the concomitant gradual banishment from all regions of human thought of what we call spirit and spontaneity.

I have endeavored, in the first part of this discourse, to give you a conception of the direction towards which modern physiology is tending ; and I ask you, what is the difference between the conception of life as the product of a certain disposition of material molecules, and the old notion of an Archæus governing and directing blind matter within each living body, except this—that here, as elsewhere, matter and law have devoured spirit and spontaneity? And as surely as every future grows out of past and present, so will the physiology of the future gradually extend the realm of matter and law until it is coëxtensive with knowledge, with feeling, and with action. The consciousnes of this great truth weighs like a nightmare, I believe, upon many of the best minds of these days. They watch what they conceive to be the progress of materialism, in such fear and powerless anger as a savage feels, when, during an eclipse, the great shadow creeps over the face of the sun. The advancing tide of matter threatens to drown their souls ; the tightening grasp of law impedes their freedom ; they are alarmed lest man's moral nature be debased by the increase of his wisdom.

If the "New Philosophy" be worthy of the reprobation with which it is visited, I confess their fears seem to me to be well founded. While, on the contrary, could

David Hume be consulted, I think he would smile at their perplexities, and chide them for doing even as the heathen, and falling down in terror before the hideous idols their own hands have raised. For, after all, what do we know of this terrible "matter," except as a name for the unknown and hypothetical cause of states of our own consciousness? And what do we know of that "spirit" over whose threatened extinction by matter a great lamentation is arising, like that which was heard at the death of Pan, except that it is also a name for an unknown and hypothetical cause, or condition, of states of consciousness? In other words, matter and spirit are but names for the imaginary substrata of groups of natural phenomena. And what is the dire necessity and " iron" law under which men groan ? Truly, most gratuitously invented bugbears. I suppose if there be an " iron" law, it is that of gravitation ; and if there be a physical necessity, it is that a stone, unsupported, must fall to the ground. But what is all we really know and can know about the latter phenomenon ? Simply, that, in all human experience, stones have fallen to the ground under these conditions ; that we have not the smallest reason for believing that any stone so circumstanced will not fall to the ground, and that we have, on the contrary, every reason to believe that it will so fall. It is very convenient to indicate that all the conditions of belief have been fulfilled in this case, by calling the statement that unsupported stones will fall to the ground, " a law of nature." But when, as commonly happens, we change will into must, we introduce an idea of necessity which most assuredly does not lie in the observed facts, and has no warranty that I can discover elsewhere. For my part, I utterly repudiate and anathematize the intruder. Fact, I know ; and Law I know ; but what is this Necessity, save an empty shadow of my own mind's throwing? But, if it is certain that we can have no knowledge of the nature of either matter or spirit, and that the notion of necessity is something illegitimately thrust into the perfectly legitimate conception of law, the materialistic position that there is nothing in the world but matter, force, and necessity, is as utterly devoid of justification as the most baseless of theological dogmas.

The fundamental doctrines of materialism, like those of spiritualism, and most other " isms," lie outside " the limits of philosophical inquiry," and David Hume's great service to humanity is his irrefragable demonstration of what these limits are. Hume called himself a sceptic, and therefore others cannot be blamed if they apply the the same title to him ; but that does not alter the fact that the name, with its existing implications, does him gross injustice. If a man asks me what the politics of the inhabitants of the moon are, and I reply that I do not know ; that neither I, nor any one else have any means of knowing ; and that, under these circumstances I decline to trouble myself about the subject at all, I do not think he has any right to call me a sceptic. On the contrary, in replying thus, I conceive that I am sim- ply honest and truthful, and show a proper regard for the economy of time. So Hume's strong and subtle in- tellect takes up a great many problems about which we are naturally curious, and shows us that they are essen- tially questions of lunar politics, in their essence incapable of being answered, and therefore not worth the attention of men who have work to do in the world. And thus ends one of his essays :

" If we take in hand any volume of Divinity, or school ·metaphysics, for instance, let us ask, *Does it contain any abstract reasoning concerning quantity or number ?* No. *Does it contain any experimental reasoning concerning matter of fact and existence ?* No. Commit it then to the flames ; for it can contain nothing but sophistry and illusion."

Permit me to enforce this most wise advice. Why trouble ourselves about matters of which, however important they may be, we do know nothing, and can know nothing? We live in a world which is full of misery and ignorance, and the plain duty of each and all of us is to try to make the little corner he can influence somewhat less miserable and somewhat less ignorant than it was before he entered it. To do this effectually it is necessary to be fully possessed of only two beliefs: the first, that the order of nature is ascertainable by our faculties to an extent which is practically unlimited ; the second, that our volition counts for something as a condition of the course of events. Each of these beliefs can be verified experimentally, as often as we like to try. Each, therefore, stands upon the strongest foundation upon which any belief can rest ; and forms one of our highest truths.

If we find that the ascertainment of the order of nature is facilitated by using one terminology, or one set of symbols, rather than another, it is our clear duty to use the former, and no harm can accrue so long as we bear in mind that we are dealing merely with terms and symbols.

In itself it is of little moment whether we express the phenomena of matter in terms of spirit, or the phenomena of spirit in terms of matter ; matter may be regarded as a form of thought, thought may be regarded as a property of matter—each statement has a certain relative truth. But with a view to the progress of science, the materialistic terminology is in every way to be preferred. For it connects thought with the other phenomena of the universe, and suggests inquiry into the nature of those physical conditions or concomitants of thought, which are more or less accessible to us, and a knowledge of which may, in future, help us to exercise the same kind of control over the world of thought as we already possess in respect of the material world ; whereas, the alternative, or spiritualistic, terminology is utterly barren, and leads to nothing but obscurity and confusion of ideas. Thus there can be little doubt that the further science advances, the more extensively and consistently will all the phenomena of nature be represented by materialistic formulæ and symbols. But the man of science, who, forgetting the limits of philosophical inquiry, slides from these formulæ and symbols into what is commonly understood by materialism, seems to me to place himself on a level with the mathematician, who should mistake the x's and y's, with which he works his problems, for real entities—and with this further disadvantage as compared with the mathematician, that the blunders of the latter are of no practical consequence, while the errors of systematic materialism may paralyze the energies and destroy the beauty of a life,

THE CORRELATION OF VITAL AND PHYSICAL FORCES.

THE CORRELATION

OF

VITAL AND PHYSICAL FORCES.

———

In the Syracusan Poecile, says Alexander von Humboldt in his beautiful little allegory of the Rhodian Genius, hung a painting, which, for full a century, had continued to attract the attention of every visitor. In the foreground of this picture a numerous company of youths and maidens of earthly and sensuous appearance gazed fixedly upon a haloed Genius who hovered in their midst. A butterfly rested upon his shoulder, and he held in his hand a flaming torch. His every lineament bespoke a celestial origin. The attempts to solve the enigma of this painting—whose origin even was unknown —though numerous, were all in vain, when one day a ship arriving from Rhodes, laden with works of art, brought another picture, at once recognized as its companion. As before, the Genius stood in the center, but the butterfly had disappeared, and the torch was reversed and extinguished. The youths and maidens were no longer sad and submissive, their mutual embraces announcing their entire emancipation from restraint. Still unable to solve the riddle, Dionysius sent the pictures to the Pythagorean sage, Epicharmus. After gazing upon them long and earnestly, he said : Sixty years long have I pondered on the internal springs of nature, and on the differences inherent in matter ; but it is only this day that the Rhodian Genius has taught me to see clearly that which before I had only conjectured. In inanimate nature, everything seeks its like. Everything, as soon as formed, hastens to enter into new combinations, and nought save the disjoining art of man can present in a separate state ingredients which ye would vainly seek in the interior of the earth or in the moving oceans of air and water. Different, however, is the blending of the same substances in animal and vegetable bodies. Here vital force imperatively asserts its rights, and heedless of the affinity and antagonism of the atoms, unites substances which in inanimate nature ever flee from each other, and separates that which is incessantly striving to unite. Recognize, therefore, in the Rhodian Genius, in the expression of his youthful vigor, in the butterfly on his shoulder, in the commanding glance of his eye, the symbol of vital force as it animates every germ of organic creation. The earthly elements at his feet are striving to gratify their own desires and to mingle with one another. Imperiously the Genius threatens them with upraised and high-flaming torch, and compels them regardless of their ancient rights, to obey his laws. Look now on the new work of art ; turn from life to death. The butterfly has soared upward, the extinguished torch is reversed, and the head of the youth is drooping ; the spirit has fled to other spheres, and the vital force is extinct. Now the youths and maidens join their hands in joyous accord. Earthly matter again resumes its rights. Released from all bonds, they impetuously follow their natural instincts, and the day of his death is to them a day of nuptials.[1]

The view here put by Humboldt into the mouth of Epicharmus may be taken as a fair representation of the current opinion of all ages concerning vital force. To-day, as truly as seventy-five years ago when Humboldt wrote, the mysterious and awful phenomena of life are commonly attributed to some controlling agent residing in the organism—to some independent presiding deity, holding it in absolute subjection. Such a notion it was which prompted Heraclitus to talk of a universal fire, Van Helmont to propose his Archæus, Hofmann his vital fluid, Hunter his *materia vitæ diffusa*, and Humboldt his vital force.[2] All these names assume the existence of a material or immaterial something, more or less separable from the material body, and more or less identical with the mind or soul, which is the cause of the phenomena of living beings. But as science moved irresistibly onward, and it became evident that the forces of inorganic nature were neither deities nor imponderable fluids, separable from matter, but were simple affections of it, analogy demanded a like concession in behalf of vital force.[3] From the notion that the effects of heat were due to an imponderable fluid called caloric, discovery passed to the conviction that heat was but a motion of material particles, and hence inseparable from matter. To a like assumption concerning vitality it was now but a step. The more advanced thinkers in science of to-day, therefore, look upon the life of the living form as inseparable from its substance, and believe that the former is purely phenomenal, and only a manifestation of the latter. Denying the existence of a special vital force as such, they retain the term only to express the sum of the phenomena of living beings.

In calling your attention this evening to the Correlation of the Physical and the Vital Forces, I have a twofold object in view. On the one hand, I would seek to interest you in a comparatively recent discovery of Science, and one which is destined to play a most important part in promoting man's welfare ; and on the other I would inquire what part our own country has had in these discoveries.

In the first place, then, let us consider what the evidences are that vital and physical forces are correlated. Let us inquire how far inorganic and organic forces may be considered mutually convertible, and hence, in so far, mutually identical. This may best be done by considering, first, what is to be understood by correlation : and second, how far are the physical forces themselves correlated to each other.

At the outset of our discussion, we are met by an unfortunate ambiguity of language. The word Force, as commonly used, has three distinct meanings ; in the first place, it is used to express the cause of motion, as when we speak of the force of gunpowder ; it is also used to indicate motion itself, as when we refer to the force of a moving cannon-ball ; and lastly it is employed to express the effect of motion, as when we speak of the blow which the moving body gives.[4] Because of this confusion, it has been found convenient to adopt Rankine's suggestion,[5] and to substitute the word 'energy' therefor. And precisely as all force upon the earth's surface—

using the term force in its widest sense—may be divided
into attraction and motion, so all energy is divided into
potential and actual energy, synonymous with those
terms. It is the chemical attraction of the atoms, or
their potential energy, which makes gunpowder so powerful ; it is the attraction or potential energy of gravitation which gives the power to a raised weight. If now,
the impediments be removed, the power just now latent
becomes active, attraction is converted into motion,
potential into actual energy, and the desired effect is
accomplished. The energy of gunpowder or of a raised
weight is potential, is capable of acting ; that of explod-
ing gunpowder or of a falling weight is actual energy
or motion. By applying a match to the gunpowder, by
cutting the string which sustains the weight, we convert
potential into actual energy. By potential energy, therefore, is meant attraction ; and by actual energy, motion.
It is in the latter sense that we shall use the word force
in this lecture ; and we shall speak of the forces of
heat, light, electricity and mechanical motion, and of
the attractions of gravitation, cohesion, chemism.

From what has now been said, it is obvious that when
we speak of the forces of heat, light, electricity or motion, we mean simply the different modes of motion
called by these names. And when we say that they
are correlated to each other, we mean simply that the
mode of motion called heat, light, electricity, is convertible into any of the others, at pleasure. Correlation
therefore implies convertibility, and mutual dependence
and relationship.

Having now defined the use of the term force, and
shown that forces are correlated which are convertible

and mutually dependent, we go on to study the evidences
of such correlation among the motions of inorganic nature usually called physical forces ; and to ask what
proof science can furnish us that mechanical motion,
heat, light, and electricity are thus mutually convertible.
As we have already hinted, the time was when these
forces were believed to be various kinds of imponderable matter, and chemists and physicists talked of the
union of iron with caloric as they talked of its union
with sulphur, regarding the caloric as much a distinct
and inconvertible entity as the iron and sulphur themselves. Gradually, however, the idea of the indestructibility of matter extended itself to force. And as it
was believed that no material particle could ever be
lost, so, it was argued, no portion of the force existing
in nature can disappear. Hence arose the idea of the
indestructibility of force. But, of course, it was quite
impossible to stop here. If force cannot be lost, the
question at once arises, what becomes of it when it
passes beyond our recognition ? This question led to
experiment, and out of experiment came the great fact
of force-correlation ; a fact which distinguished authority
has pronounced the most important discovery of the
present century.[6] These experiments distinctly proved
that when any one of these forces disappeared, another
took its place ; that when motion was arrested, for example, heat, light or electricity was developed. In short,
that these forces were so intimately related or correlated
—to use the word then proposed by Mr. Grove[7]—that
when one of them vanished, it did so only to reappear
in terms of another. But one step more was necessary
to complete this magnificent theory. What can produce

motion but motion itself? Into what can motion be converted, but motion? May not these forces, thus mutually convertible, be simply different modes of motion of
the molecules of matter, precisely as mechanical motion
is a motion of its mass? Thus was born the dynamic
theory of force, first brought out in any completeness by
Mr. Grove, in 1842, in a lecture on the " Progress of
Physical Science," delivered at the London Institution.
In that lecture he said : " Light, heat, electricity, magnetism, motion, are all convertible material affections.
Assuming either as the cause, one of the others will be
the effect. Thus heat may be said to produce electricity,
electricity to produce heat ; magnetism to produce electricity, electricity magnetism ; and so of the rest." [8]

A few simple experiments will help us to fix in our
minds the great fact of the convertibility of force.
Starting with actual visible motion, correlation requires
that when it disappears as motion, it should reappear as
heat, light, or electricity. If the moving body be elastic
like this rubber ball, then its motion is not destroyed
when it strikes, but is only changed in direction. But
if it be non-elastic, like this ball of lead, then it does
not rebound ; its motion is converted into heat. The
motion of this sledge-hammer, for example, which if received upon this anvil would be simply changed in
direction, if allowed to fall upon this bar of lead, is
converted into heat ; the evidence of which is that a
piece of phosphorus placed upon the lead is at once in-
flamed. So too, if motion be arrested by the cushion
of air in this cylinder, the heat evolved fires the tinder
carried in the plunger. But it is not necessary that the
arrest of motion should be sudden ; it may be gradual,

as in the case of friction. If this cylinder containing water or alcohol be caused to revolve rapidly between the two sides of this wooden rubber, the heat due to the arrested motion will raise the temperature of the liquid to the boiling point, and the cork will be expelled. But motion may also be converted into electricity. Indeed electricity is always the result of friction between heterogeneous particles.[9] When this piece of hard rubber, for example, is rubbed with the fur of a cat, it is at once electrified ; and now if it be caused to communicate a portion of its charge to this glass plate, to which at the same time we add the mechanical motion of rotation, the strong sparks produced give evidence of the conversion.

So, too, taking heat as the initial force, motion, light, electricity may be produced. In every steam-engine the steam which leaves the cylinder is cooler than that which entered it, and cooler by exactly the amount of work done. The motion of the piston's mass is precisely that lost by the steam molecules which batter against it. The conversion of heat into electricity, too, is also easily effected. When the junction of two metals is heated, electricity is developed. If the two metals be bismuth and antimony, as represented in this diagram, the currents flow as indicated by the arrows ; and by multiplying the number of pairs, the effect may be proportionately increased. Such an arrangement, called a thermo-electric battery, we have here ; and by it the heat of a single gas-burner may be made to move, when converted, this little electric bell-engine. Moreover, heat and light have the very closest analogy ; exalt the rapidity with which the molecules move and light appears, the difference being only one of intensity.

Again, if electricity be our starting point, we may accomplish its conversion into the other forces. Heat
results whenever its passage is interrupted or resisted ;
a wire of the poorly conducting metal platinum becoming even red hot by the converted electricity. To produce light, of course, we need only to intensify this
action ; the brightest artificial light known, results from
a direct conversion of electricity.

Enough has now been said to establish our point.
What is to be particularly observed of these pieces
of apparatus is that they are machines especially designed for the conversion of some one force into another. And we expect of them only that conversion.
We pass on to consider for a moment the quantitative relations of this mutual convertibility. We notice, in the first place, that in all cases save one, the
conversion is not perfect, a part of the force used not
being utilized, on the one hand, and on the other,
other forces making their appearance simultaneously.
While, for example, the conversion of motion into heat
is quite complete, the inverse conversion is not at all so.
And on the other hand, when motion is converted into
electricity, a part of it appears as heat. This simultaneous production of many forces is well illustrated by
our little bell-engine, which converts the electricity of
the thermo-battery into magnetism, and this into motion,
a part of which expends itself as sound. For these
reasons the question " How much ?" is one not easily
answered in all cases. The best known of these relations is that between motion and heat, which was first
established by Mr. Joule in 1849, after seven years of
patient investigation.[10] The apparatus which he used is

shown in the diagram. It consists of a cylindrical box of metal, through the cover of which passes a shaft, carrying upon its lower end a set of paddles, immersed in water within the box, and upon its upper portion a drum, on which are wound two cords, which, passing in opposite directions, run over pulleys, and are attached to known weights. The temperature of the water within the box being carefully noted, the weights are then allowed to fall a certain number of times, of course in their fall turning the paddles against the friction of the liquid. At the close of the experiment the water is found to be warmer than before. And by measuring the amount of this rise in temperature, knowing the distance through which the weights have fallen, it is easy to calculate the quantity of heat which corresponds to a given amount of motion. In this way, and as a mean of a large number of experiments, Mr. Joule found that the amount of mass motion in a body weighing one pound, which had fallen from a hight of 772 feet, was exactly equal to the molecular motion which must be added to a pound of water, in order to heat it one degree Fahrenheit. If we call the actual energy of a body weighing one pound which has fallen one foot, a foot-pound, then we may speak of the mechanical equivalent of heat as being 772 foot-pounds.

The significance and value of this numerical constant will appear more clearly if we apply it to the solution of one or two simple problems. During the recent war two immense iron guns were cast in Pittsburgh, whose weight was nearly 112,000 pounds each, and which had a caliber of 20 inches.[11] Upon this diagram is a calculation of the effective blow which the solid shot of such a gun, assum-

ing its weight to be 1,000 pounds and its velocity 1,100
feet per second, would give ; it is 902,797 tons !¹² Now,
if it were possible to convert the whole of this enormous
mechanical power into heat, to how much would it correspond ? This question may be answered by the aid
of the mechanical equivalent of heat ; here is the calculation, from which we see that when 17 gallons of
ice-cold water are heated to the boiling point, as much
energy is communicated as is contained in the death-
dealing missile at its highest velocity.¹³ Again, if we take
the impact of a larger cannon-ball, our earth, which is
whirling through space with a velocity of 19 miles a
second, we find it to be 98,416,136,000,000,000,000,000,-
000,000,000 tons !¹⁴ Were this energy all converted into
heat, it would equal that produced by the combustion
of 14 earths of solid coal.¹⁵

The conversion of heat into motion, however, as already stated, is not as perfect. The best steam-engines
economize only one-twentieth of the heat of the fuel.¹⁶
Hence if a steamship require 600 tons of coal to carry
her across the Atlantic, 570 tons will be expended in
heating the waters of the ocean, the heat of the remaining 30 tons only being converted into work.

One other quantitative determination of force has
also been made. Prof. Julius Thomsen, of Copenhagen,
has fixed experimentally the mechanical equivalent of
light.¹⁷ He finds that the energy of the light of a sper-
maceti candle burning 126½ grains per hour, is equal
in mechanical value to 13·1 foot-pounds per minute.
The same conclusion has been reached by Mr. Farmer,
of Boston, from different data.¹⁸

If we pass from the actual physical energies or motions to consider for a moment the potential energies or attractions, we find, also, an intimate correlation. Since all energy not active in motion is potential in attraction, it follows that in the attractions we have energy stored up for subsequent use. The sun is thus storing up energy : every minute it raises 2,000,000,000 tons of water to the mean hight of the clouds, 3½ miles ; and the actual energy set free when this water falls is equal to 2,757,000,000,000 horse powers.[19]   So when the oxygen and the zinc of the ore are separated in the furnace, the actual energy of heat becomes the potential energy of chemical .attraction, which again becomes actual in the form of electricity when the zinc is dissolved in an acid. We see, then, that not only may any form of force or actual energy be stored up as any form of attraction or potential energy, but that the latter, from whatsoever source derived, may appear as heat, light, electricity, or mechanical motion.

Having now established the fact of correlation for the physical forces, we have next to inquire what are the evidences of the correlation of the vital forces with them.   But in the first place it must be remarked that life is not a simple term like heat or electricity ; it is a complex term, and includes all those phenomena which a living body exhibits.   In this discussion, therefore, we shall use the term vital force to express only the actual energy of the body, however manifested.   As to the attractions or the potential energy of the organism, nothing is more fully settled in science than the fact that these are precisely the same within the body as without it.   Every particle of matter within the body obeys implicitly the laws of the chemical and physical attractions. No overpowering or supernatural agency comes in to complicate their action, which is modified only by the action of the others. Vitality, therefore, is the sum of the energies of a living body, both potential and actual.

Moreover, the important fact must be fully recognized that in living beings we have to do with no new elementary forms of matter. Precisely the same atoms which build up the inorganic fabric, compose the organic. In the early days of chemistry, indeed, it was supposed that the complicated molecules which life produced were beyond the reach of simple chemical law. But as more and more complex molecules have been, one after another, produced, chemistry has become re-assured, and now doubts not her ability to produce them all. A few years hence, and she will doubtless give us quinine and protagon, as she now gives us coumarin and neurine, substances the synthesis of which was but yesterday an impossibility.[20]

In studying the phenomena of living beings, it is important also to bear in mind the different and at the same time the coördinate purposes subserved by the two great kingdoms of nature. The food of the plant is matter whose energy is all expended; it is a fallen weight. But the plant-organism receives it, exposes it to the sun's ray, and, in a way yet mysterious to us, converts the actual energy of the sunlight into potential energy within it. The fallen weight is thus raised, and energy is stored up in substances which now are alone competent to become the food of the animal. This food is not such because any new atoms have been added to it ; it is food because it contains within it potential energy, which at any time may become actual as force. This food the animal now appropriates ; he brings it in contact with oxygen, and the potential energy becomes actual ; he cuts the string, the weight falls, and what was just now only attraction, has become actual force ; this force he uses for his own purposes, and hands back the oxidized matter, the fallen weight, to the plant to be again de-oxidized, to be again raised. The plant then is to regarded as a machine for converting sunlight into potential energy ; the animal, a machine for setting the potential energy free as actual, and economizing it. The force which the plant stores up is undeniably physical ; must not the force which the animal sets free by its conversion, be intimately correlated to it?

But approaching our question still more closely, let us, in illustration of the vital forces of the animal economy, choose three forms of its manifestation in which to seek for the evidences of correlation ; these shall be heat, evolved within the body ; muscular energy or motion ; and lastly, nervous energy, or that form of force which, on the one hand, stimulates a muscle to contract, and on the other, appears in forms called mental.

The heat which is produced by the living body is obviously of the same nature as heat from any other source ; it is recognized by the same tests, and may be applied for the same purposes. As to its origin, it is evident that since potential energy exists in the food which enters the body, and is there converted into force, a portion of it may become the actual energy of heat. And since, too, the heat produced in the body is precisely such as would be set free by the combustion of this food outside of it, it is fair to assume that it thus originates. To this may be added the chemical argument that while food capable of yielding heat by combustion is taken into the body, its constituents are completely or almost completely, oxidized before leaving it ; and since oxidation always evolves heat, the heat of the body must have its origin in the oxidation of the food. Moreover, careful measurements have demonstrated that the amount of heat given off by the body of a man weighing 180 pounds is about 2,500,000 units. Accurate calculations have shown, on the other hand, that 288·4 grams of carbon and 12·56 grams of hydrogen are available in the daily food for the production of heat. If burned out of the body, these quantities of carbon and hydrogen would yield 2,765,134 heat units. Burned within it, as we have just seen, 2,500,000 units appear as heat ; the rest in other forms of energy.[21] We conceive, however, that no long argument is necessary to prove that animal heat results from a conversion of energy within the body ; or that the vital force heat, is as truly correlated to the other forces as when it it has a purely physical origin.

The belief that the muscular force exerted by an animal is created by him is by no means confined to the very earliest ages of history. Traces of it appear to the careful observer even now, although, as Dr. Frankland says, science has proved that "an animal can no more generate an amount of force capable of moving a grain of sand than a stone can fall upward or a locomotive drive a train without fuel."[22] In studying the characters of muscular action we notice, first, that, as in the case of heat, the force which it develops is in no wise different from motion in inorganic nature. In the early part of the lecture, motion produced by the contraction of muscle, was used to show the conversion of mass-force into molecular force. No one in this room believes, I presume, that the result would have been at all different, had the motion been supplied by a steam-engine or a water-wheel. Again, food, as we have seen, is of value for the potential energy it contains, which may become actual in the body. Liebig, in 1842, asserted that for the production of muscular force, the food must first be converted into muscular tissue,[23] a view until recently accepted by physiologists.[24] It has been conclusively shown, however, within a few years, that muscular force cannot come from the oxidation of its own substance, since the products of this metamorphosis are not increased in amount by muscular exertion.[25] Indeed, reasoning from the whole amount of such products excreted, the oxidation of the amount of muscle which they represent would furnish scarcely one-fifth of the mechanical force of the body. But while the products of tissue-oxidation do not increase with the increase of muscular exertion, the amount of carbonic gas exhaled by the lungs is increased in the exact ratio of the work done.[26] No doubt can be entertained, therefore, that the actual energy of the muscle is simply the converted potential energy of the carbon of the food. A muscle, therefore, like a steam-engine, is a machine for converting the potential energy of carbon into motion. But unlike a steam-engine, the muscle accomplishes this conversion directly, the energy not passing through the intermediate stage of heat. For this reason, the muscle is the most economical producer of mechanical force known. While no machine whatever can transform all of the energy into motion—the most economical steam engines utilizing only one-twentieth of the heat—the muscle is able to convert one-fifth of the energy of the food into work.[27]   The other four-fifths must, therefore, appear as heat.   Whenever a muscle contracts, then, four times as much energy appears as heat as is converted into motion.   Direct experiments by Heidenhain have confirmed this, by showing that an important rise of temperature attends muscular contraction ; [28] a fact, however, apparent to any one who has ever taken active exercise.   The work done by the animal body is of two sorts, internal and external.   The former includes the action of the heart, of the respiratory muscles, and of those assisting the digestive process.   The latter refers to the useful work the body may perform.   Careful estimates place the entire work of the body at about 800 foot-tons daily ; of which 450 foot-tons is internal, 350 foot-tons external work.   And since the internal work ultimately appears as heat within the body, the actual loss of heat by the production of motion is the equivalent of the 350 foot-tons which represents external work.   This by a simple calculation will be found to be 250,000 heat units, almost the precise amount by which the heat yielded by the food when burned without the body, exceeds that actually evolved by the organism. Moreover, while the total heat given off by the body is 2,500,000 units, the amount of energy evolved as work is equal to about 600,000 heat units ; hence the amount of work done by a muscle is as above stated, one-fifth of the actual energy derivable from the food.   One point further. The law of correlation requires that the heat set free when a muscle in contracting does work, shall be less than when it effects nothing ; this fact, too, has been

experimentally established by Heidenhain.[29]   So, again,
when muscular contraction does not result in motion,
as when one tries to raise a weight too heavy for him,
the energy which would have appeared as work, takes
the form of heat: a result deducible by the law of correlation from the steam-engine.

The last of the so-called vital forces which we are to
examine, is that produced by the nerves and nervous
centers.   In the nerve which stimulates a muscle to
contract, this force is undeniably motion, since it is
propagated along this nerve from one extremity to the
other.   In common language, too, this idea finds cur-
rency in the comparison of this force to electricity ; the
gray or cellular matter being the battery, the white or
fibrous matter the conductors.   That this force is not
electricity, however, Du Bois-Reymond has demonstrated
by showing that its velocity is only 97 feet in a second,
a speed equaled by the greyhound and the race-horse.[30]
In his opinion, the propagation of a nervous impulse is
a sort of successive molecular polarization, like magnetism.   But that this agent is a force, as analogous to
electricity as is magnetism, is shown not only by the
fact that the transmission of electricity along a nerve
will cause the contraction of the muscle to which it
leads, but also by the more important fact that the contraction of a muscle is excited by diminishing its normal electrical current ;[31] a result which could take place
only with a stimulus closely allied to electricity.   Nerve-
force, therefore, must be a transmuted potential energy.

What, now, shall we say of that highest manifestation
of animal life, thought-power ?   Has the upper region
called intelligence and reason, any relations to physical

force ? This realm has not escaped the searching investigation of modern science ; and although in it investigations are vastly more difficult than in any of the
regions thus far considered, yet some results of great
value have been obtained, which may help us to a solution of our problem. It is to be observed at the outset
that every external manifestation, of thought-force is a
muscular one, as a word spoken or written, a gesture, or
an expression of the face ; and hence this force must
be intimately correlated with nerve-force. These manifestations, reaching the mind through the avenues of
sense, awaken accordant trains of thought only when
this muscular evidence is understood. A blank sheet
of paper excites no emotion ; even covered with Assyrian cuneiform characters, its alternations of black and
white awaken no response in the ordinary brain. It is
only when, by a frequent repetition of these impressions,
the brain-cell has been educated, that these before
meaningless characters awaken thought. Is thought,
then, simply a cell action which may or may not result
in muscular expression—an action which originates new
combinations of truth only, precisely as a calculating
machine evolves new combinations of figures ? Whatever we define thought to be, this fact appears certain,
that it is capable of external manifestation by conversion into the actual energy of motion, and only by this
conversion. But here the question arises, Can it be
manifested inwardly without such a transformation of
energy ? Or is the evolution of thought entirely independent of the matter of the brain ? Experiments, ingenious and reliable, have answered this question. The
importance of the results will, I trust, warrant me in

examining the methods employed in these experiments somewhat in detail. Inasmuch as our methods for measuring minute amounts of electricity are very perfect, and the methods for the conversion of heat into electricity are equally delicate, it has been found that smaller differences of temperature may be recognized by converting the heat into electricity, than can be detected thermometrically. The apparatus, first used by Melloni in 1832,[32] is very simple, consisting first, of a pair of metallic bars like those described in the early part of the lecture, for effecting the conversion of the heat ; and second, of a delicate galvanometer, for measuring the electricity produced. In the experiments in question one of the bars used was made of bismuth, the other of an alloy of antimony and zinc.[33] Preliminary trials having shown that any change of temperature within the skull was soonest manifested externally in that depression which exists just above the occipital protuberance, a pair of these little bars was fastened to the head at this point ; and to neutralize the results of a general rise of temperature over the whole body, a second pair, reversed in direction, was attached to the leg or arm, so that if a like increase of heat came to both, the electricity developed by one would be neutralized by the other, and no effect be produced upon the needle unless only one was affected. By long practice it was ascertained that a state of mental torpor could be induced, lasting for hours, in which the needle remained stationary. But let a person knock on the door outside the room, or speak a single word, even though the experimenter remained absolutely passive, and the reception of the intelligence caused the needle to swing

through 20 degrees.[34]   In explanation of this production
of heat, the analogy of the muscle at once suggests
itself. No conversion of energy is complete ; and as
the heat of muscular action represents force which has
escaped conversion into motion, so the heat evolved
during the reception of an idea, is energy which has escaped conversion into thought, from precisely the same
cause.   Moreover, these experiments have shown that
ideas which affect the emotions, produce most heat in
their reception ; " a few minutes' recitation to one's self
of emotional-poetry, producing more effect than several
hours of deep thought." Hence it is evident that the
mechanism for the production of deep thought, accomplishes this conversion of energy far more perfectly
than that which produces simply emotion.   But we may
take a step further in this same direction.   A muscle,
precisely as the law of correlation requires, develops
less heat when doing work than when it contracts without doing it.   Suppose, now, that beside the simple reception of an idea by the brain, the thought is expressed
outwardly by some muscular sign.   The conversion now
takes two directions, and in addition to the production
of thought, a portion of the energy appears as nerve and
muscle-power ; less, therefore, should appear as heat,
according to our law of correlation.   Dr. Lombard's experiments have shown that the amount of heat developed by the recitation to one's self of emotional poetry,
was in every case less when that recitation was oral ;
i. e., had a muscular expression.   These results are in
accordance with the well-known fact that emotion often
finds relief in physical demonstrations ; thus diminishing
the emotional energy by converting it into muscular.

Nor do these facts rest upon physical evidence alone. Chemistry teaches that thought-force, like muscle-force, comes from the food ; and demonstrates that the force evolved by the brain, like that produced by the muscle, comes not from the disintegration of its own tissue, but is the converted energy of burning carbon.[35] Can we longer doubt, then, that the brain, too, is a machine for the conversion of energy?   Can we longer refuse to believe that even thought is, in some mysterious way, correlated to the other natural forces? and this, even in face of the fact that it has never yet been measured ? [36]

I cannot close without saying a word concerning the part which our own country has had in the development of these great truths.   Beginning with heat, we find that the material theory of caloric is indebted for its overthrow more to the distinguished Count Rumford than to any other one man.   While superintending the boring of cannon at the Munich Arsenal towards the close of the last century, he was struck by the large amount of heat developed, and instituted a careful series of experiments to ascertain its origin.   These experiments led him to the conclusion that " anything which any insulated body or system of bodies can continue to furnish without limitation, cannot possibly be a material substance."   But this man, to whom must be ascribed the discovery of the first great law of the correlation of energy, was an American.   Born in Woburn, Mass., in 1753, he, under the name of Benjamin Thompson, taught school afterward at Concord, N. H., then called Rumford.   Unjustly suspected of toryism during our Revolutionary war, he went abroad and distinguished himself in the service of several of the Governments of

Europe. He did not forget his native land, though she had treated him so unfairly ; when the honor of knighthood was tendered him, he chose as his title the name of the Yankee village where he had taught school, and was thenceforward known as Count Rumford. And at his death, by founding a professorship in Harvard College, and donating a prize-fund to the American Academy of Arts and Sciences at Boston, he showed his interest in her prosperity and advancement.[37] Nor has the field of vital forces been without earnest workers belonging to our own country. Professors John W. Draper[38] and Joseph Henry[39] were among its earliest explorers. And in 1851, Dr. J. H. Watters, now of St. Louis, published a theory of the origin of vital force, almost identical with that for which Dr. Carpenter, of London, has of late received so much credit. Indeed, there is some reason to believe that Dr. Watters's essay may have suggested to the distinguished English physiologist the germs of his own theory.[40] A paper on this subject by Prof. Joseph Leconte, of Columbia, S. C., published in 1859, attracted much attention abroad.[41] The remarkable results already given on the relation of heat to mental work, which thus far are unique in science, we owe to Professor J. S. Lombard, of Harvard College ; [42] the very combination of metals used in his apparatus being devised by our distinguished electrical engineer, Mr. Moses G. Farmer. Finally, researches conducted by Dr. T. R. Noyes in the Physiological Laboratory of Yale College, have confirmed the theory that muscular tissue does not wear during action, up to the point of fatigue ; [43] and other researches by Dr. L. H. Wood have first established the same great truth for brain-tissue.[44]

222324222

We need not be ashamed, then, of our part in this advance in science. Our workers are, indeed, but few; but both they and their results will live in the records of the world's progress. More would there be now of them were such studies more fostered and encouraged. Self-denying, earnest men are ready to give themselves up to the solution of these problems, if only the means of a bare subsistence be allowed them. When wealth shall foster science, science will increase wealth—wealth pecuniary, it is true: but also wealth of knowledge, which is far better.

In looking back over the whole of this discussion, I trust that it is possible to see that the objects which we had in view at its commencement have been more or less fully attained. I would fain believe that we now see more clearly the beautiful harmonies of bounteous nature; that on her many-stringed instrument force answers to force, like the notes of a great symphony; disappearing now in potential energy, and anon reappearing as actual energy, in a multitude of forms. I would hope that this wonderful unity and mutual interaction of force in the dead forms of inorganic nature, appears to you identical in the living forms of animal and vegetable life, which make of our earth an Eden. That even that mysterious, and in many aspects awful, power of thought, by which man influences the present and future ages, is a part of this great ocean of energy. But here the great question rolls upon us, Is it only this? Is there not behind this material substance, a higher than molecular power in the thoughts which are immortalized in the poetry of a Milton or a Shakespeare, the art creations of a Michael Angelo or a Titian, the har-

monies of a Mozart or a Beethoven? Is there really
no immortal portion separable from this brain-tissue,
though yet mysteriously united to it? In a word, does
this curiously-fashioned body inclose a soul, God-given
and to God returning? Here Science veils her face
and bows in reverence before the Almighty. We have
passed the boundaries by which physical science is in-
closed. No crucible, no subtle magnetic needle can
answer now our questions. No word but His who
formed us, can break the awful silence. In presence of
such a revelation Science is dumb, and faith comes in
joyfully to accept that higher truth which can never be
the object of physical demonstration.

1 HUMBOLDT, Views of Nature, Bohn's ed., London, 1850, p. 380. This allegory did not appear in the first edition of the Views of Nature. In the preface to the second edition the author gives the following account of its origin : "Schiller," he says, "in remembrance of his youthful medical studies, loved to converse with me, during my long stay at Jena, on physiological subjects." * * * "It was at this period that I wrote the little allegory on Vital Force, called The Rhodian Genius. The predilection which Schiller entertained for this piece, which he admitted into his periodical, *Die Horen*, gave me courage to introduce it here." It was published in *Die Horen* in 1795.

2 HUMBOLDT, *op. cit.*, p. 386. In his *Aphorismi ex doctrina Physiologiæ chemicæ Plantarum*, appended to his *Flora Fribergensis subterranea*, published in 1793, Humboldt had said "Vim internam, quæ chymicæ affinitatis vincula resolvit, atque obstat, quominus elementa corporum libere conjungantur, vitalem vocamus." "That internal force, which dissolves the bonds of chemical affinity, and prevents the elements of bodies from freely uniting, we call vital." But in a note to the allegory above mentioned, added to the third edition of the Views of Nature in 1849, he says : "Reflection and prolonged study in the departments of physiology and chemistry have deeply shaken my earlier belief in peculiar so-called vital forces. In the year 1797, * * * I already declared that I by no means regarded the existence of these peculiar vital forces as established." And again : "The difficulty of satisfactorily referring the vital phe-

nomena of the organism to physical and chemical laws depends chiefly (and almost in the same manner as the prediction of meteorological processes in the atmosphere) on the complication of the phenomena, and on the great number of the simultaneously acting forces as well as the conditions of their activity."

3 Compare HENRY BENCE JONES, Croonian Lectures on Matter and Force. London, 1868, John Churchill & Sons.

4 Ib., Preface, p. vi.

5 RANKINE, W. J. M., Philosophical Magazine, Feb., 1853. Also Edinburgh Philosophical Journal, July, 1855.

6 ARMSTRONG, Sir WM. In his address as President of the British Association for the Advancement of Science. Rep. Brit. Assoc., 1863, li.

7 GROVE, W. R., in 1842. Compare "Nature" i, 335, Jan. 27, 1870. Also Appleton's Journal, iii, 324, Mch. 19, 1870.

8 Id., in Preface to The Correlation of Physical Forces, 4th ed. Reprinted in The Correlation and Conservation of Forces, edited by E. L. Youmans, p. 7. New York, 1865, D. Appleton & Co.

9 Id., ib., Am. ed., p. 33 et seq.

10 JOULE, J. P., Philosophical Transactions, 1850, p. 61.

11 See American Journal of Science, II, xxxvii, 296, 1864.

12 The work (W) done by a moving body is commonly expressed by the formula $W = MV^2$, in which M, or the mass of the body, is equal to $\frac{w}{2g}$; *i. e.*, to the weight divided by twice the intensity of gravity. The work done by our cannon-ball then, would be $\frac{1 \times (1100)^2}{2 \times 64\frac{1}{3}} = 9{,}404 \cdot 14$ foot-tons. If, further, we assume the resisting body to be of such a character as to bring the ball to rest in moving $\frac{1}{4}$ of an inch, then the final pressure would be $9{,}404 \cdot 14 \times 12 \times 4 = 451{,}398 \cdot 7$ tons. But since, "in the case of a perfectly elastic body, or of a resistance proportional to the advance of the center of gravity of the impinging body from the point at which contact first takes place, the final pressure (provided the body

struck is perfectly rigid) is double what would occur were the stop-
page to occur at the end of a corresponding advance against a uniform resistance," this result must be multiplied by two; and we get
(451,398·7×2) 902,797 tons as the crushing pressure of the ball under these conditions. [The author's thanks are due to his friends
Pres. F. A. P. Barnard,and Mr. J. J. Skinner for suggestions on
the relation of impact to statical pressure.]

13 The unit of impact being that given by a body weighing one
pound and moving one foot a second, the impact of such a body
falling from a hight of 772 feet—the velocity acquired being $222\frac{1}{4}$
feet per second $(=\sqrt{2sg})$—would be $1\times(222\frac{1}{4})^2=49,408$ units, the
equivalent in impact of one heat-unit. A cannon-ball weighing
1000 lbs. and moving 1100 feet a second would have an impact of .
$(1100)^2\times1000=1,210,000,000$ units. Dividing this by 49,408, the
quotient is 24489 heat-units, the equivalent of the impact. The
specific heat of iron being ·1138, this amount of heat would raise
the temperature of one pound of iron 215,191° F. (24,489×·1138) or
of 1000 pounds of iron 215° F. 24489 pounds of water heated one
degree, is equal to $136\frac{1}{2}$ pounds, or 17 gallons U. S., heated 180
degrees; i. e., from 32° to 212° F.

14 Assuming the density of the earth to be 5·5, its weight would
be 6,500,000,000,000,000,000,000 tons, and its impact—by the formula given above—would be 1,025,000,000,000,000,000,000,000,-
000 foot-tons. Making the same supposition as in the case of our
cannon-ball, the final pressure would be that here stated.

15 TYNDALL, J., Heat considered as a mode of Motion; Am. ed.,
p. 57, New York, 1863.

16 RANKINE (The Steam-engine and other prime Movers, London, 1866,) gives the efficiency of Steam-engines as from 1-15th to
1-20th of the heat of the fuel.

ARMSTRONG, Sir WM., places this efficiency at 1-10th as the
maximum. In practice, the average result is only 1-30th. Rep.
Brit. Assoc., 1863, p. liv.

HELMHOLTZ, H. L. F., says : "The best expansive engines give
back as mechanical work only eighteen per cent. of the heat generated by the fuel." Interaction of Natural Forces, in Correlation
and Conservation of Forces, p. 227.

17 THOMSEN, JULIUS, Poggendorff's Annalen, cxxv, 348 Also in abstract in Am. J. Sci., II, xli, 396, May, 1866.

18 American Journal of Science, II, xli, 214, March, 1856.

19 In this calculation the annual evaporation from the ocean is assumed to be about 9 feet. (See Dr. BUIST, quoted in Maury's Phys. Geography of the Sea, New York, 1861, p. 11.) Calling the water-area of our globe 150,000,000 square miles, the total evaporation in tons per minute, would be that here given. Inasmuch as 30,000 pounds raised one-foot high is a horse-power, the number of horse-powers necessary to raise this quantity of water $3\frac{1}{2}$ miles in one minute is 2,757,000,000,000. This amount of energy is precisely that set free again when this water falls as rain.

20 Compare ODLING, WM., Lectures on Animal Chemistry, London, 1866. "In broad antagonism to the doctrines which only a few years back were regarded as indisputable, we now find that the chemist, like the plant, is capable of producing from carbonic acid and water a whole host of organic bodies, and we see no reason to question his ultimate ability to reproduce all animal and vegetable principles whatsoever." (p. 52.)

"Already hundreds of organic principles have been built up from their constituent elements, and there is now no reason to doubt our capability of producing all organic principles whatsoever in a similar manner." (p. 58.)

Dr. Odling is the successor of Faraday as Fullerian Professor of Chemistry in the Royal Institution of Great Britain.

21 MARSHALL, JOHN, Outlines of Physiology, American edition, 1868, p. 916.

22 FRANKLAND, EDWARD, On the Source of Muscular Power, Proc. Roy. Inst., June 8, 1866 ; Am. J. Sci., II, xlii, 393, Nov. 1866.

23 LIEBIG, JUSTUS VON, Die organische Chemie in ihrer Anwendung auf Physiologie und Pathologie, Braunschweig, 1842. Also in his Animal Chemistry, edition of 1852 (Am. ed., p. 26), where he says "Every motion increases the amount of organized tissue which undergoes metamorphosis."

24 Compare DRAPER, JOHN WM. Human Physiology.

PLAYFAIR, LYON, On the Food of Man in relation to his useful ᴡ..ᴋ, Edinburgh, 1865. Proc. Roy. Inst., Apr. 28, 1865.
RANKE, Tetanus eine Physiologische Studie, Leipzig, 1865.
ODLING, *op. cit.*

25 VOIT, E., Untersuchungen über den Einfluss des Kochsalzes, des Kaffees, und der Muskelbewegungen auf den Stoffwechsel, Munich, 1860.
SMITH, E., Philosophical Transactions, 1861, 747.
FICK, A., and WISLICENUS, J., Phil. Mag., IV, xxxi, 485.
FRANKLAND, E., *loc. cit.*
NOYES, T. R., American Journal Medical Sciences, Oct. 1867.
PARKES, E. A., Proceedings Royal Society, xv, 339 ; xvi, 44.

26 SMITH, EDWARD, Philosophical Transactions, 1859, 709.

27 Authorities differ as to the amount of energy converted by the steam-engine. (See Note 16.) Compare MARSHALL. *op. cit.*, p. 918. "Whilst, therefore, in an engine one-twentieth part only of the fuel consumed is utilized as mechanical power, one-fifth of the food absorbed by man is so appropriated."

28 HEIDENHAIN, Mechanische Leistung Wärmeentwickelung und Stoffumsatz bei der Muskelthätigkeit, Breslau, 1864.
See also HAUGHTON, SAMUEL, On the Relation of Food to work, published in "Medicine in Modern Times," London, 1869, Macmillan & Co.

29 HEIDENHAIN, *op. cit.* Also by FICK, Untersuchungen über Muskel-arbeit, Basel, 1867. Compare also "Nature," i, 159, Dec. 9, 1869.

30 DU BOIS-REYMOND, EMIL, On the time required for the transmission of volition and sensation through the nerves, Proc. Roy. Inst. Also in Appendix to Bence Jones's Croonian lectures.

31 MARSHALL, *op. cit.*, p. 227.

32 MELLONI, Ann. Ch. Phys., xlviii, 198.
See also NOBILI, Bibl. Univ., xliv, 225, 1830; lvii, 1, 1834.

33 The apparatus employed is illustrated and fully described in Brown-Sequard's Archives de Physiologie, i, 498, June, 1868. By it the 1-4000th of a degree Centigrade may be indicated.

34 LOMBARD, J. S., New York Medical Journal, v, 198, June, 1867. [A part of these facts were communicated to me directly by th.'r discoverer.]

35 WOOD, L. H., On the influence of Mental activity on the Excretion of Phosphoric acid by the Kidneys. Proceedings Connecticut Medical Society for 1869, p. 197.

36 On this question of vital force, see LIEBIG, Animal Chemistry. " The increase of mass in a plant is determined by the occurrence of a decomposition which takes place in certain parts of the plant under the influence of light and heat."

" The modern science of Physiology has left the track of Aristotle. To the eternal advantage of science, and to the benefit of mankind it no longer invents a *horror vacui*, a *quinta essentia*, in order to furnish credulous hearers with solutions and explanations of phenomena, whose true connection with others, whose ultimate cause is still unknown."

" All the parts of the animal body are produced from a peculiar fluid circulating in its organism, by virtue of an influence residing in every cell, in every organ, or part of an organ."

" Physiology has sufficiently decisive grounds for the opinion that every motion, every manifestation of force, is the result of a transformation of the structure or of its substance; that every conception, every mental affection, is followed by changes in the chemical nature of the secreted fluids; that every thought, every sensation is accompanied by a change in the composition of the substance of the brain."

" All vital activity arises from the mutual action of the oxygen of the atmosphere and the elements of the food."

" As, in the closed galvanic circuit, in consequence of certain changes which an inorganic body, a metal, undergoes when placed in contact with an acid, a certain something becomes cognizable by our senses, which we call a current of electricity; so in the animal body, in consequence of transformations and changes undergone by matter previously constituting a part of the organism, certain phenomena of motion and activity are perceived, and these we call life, or vitality."

" In the animal body we recognize as the ultimate cause of all force only one cause, the chemical action which the elements of the food and the oxygen of the air mutually exercise on each other. The only known ultimate cause of vital force, either in animals or in plants, is a chemical process."

" If we consider the force which determines the vital phenomena as a property of certain substances, this view leads of itself to a new and more rigorous consideration of certain singular phenomena, which these very substances exhibit, in circumstances in which they no longer make a part of living organisms."

Also OWEN, RICHARD, (Derivative Hypothesis of Life and Species, forming the 40th chapter of his Anatomy of Vertebrates, republished in Am. J. Sci., II, xlvii, 33, Jan. 1869.) " In the endeavor to clearly comprehend and explain the functions of the combination of forces called 'brain,' the physiologist is hindered and troubled by the views of the nature of those cerebral forces which the needs of dogmatic theology have imposed on mankind." * * " Religion pure and undefiled, can best answer how far it is righteous or just to charge a neighbor with being unsound in his principles who holds the term 'life' to be a sound expressing the sum of living phenomena; and who maintains these phenomena to be modes of force into which other forms of force have passed, from potential to active states, and reciprocally, through the agency of these sums or combinations of forces impressing the mind with the ideas signified by the terms 'monad,' 'moss,' 'plant,' or 'animal.' "

And HUXLEY, THOS. H., " On the Physical Basis of Life," University Series, No. 1. College Courant, 1870.

Per contra, see the Address of Dr. F. A. P. Barnard, as retiring President, before the Am. Assoc. for the Advancement of Science, Chicago meeting, August, 1868. " Thought cannot be a physical force, because thought admits of no measure."

GOULD, BENJ. APTHORP, Address as retiring President, before the American Association at its Salem meeting, Aug., 1869.

BEALE, LIONEL S., " Protoplasm, or Life, Matter, and Mind." London, 1870. John Churchill & Sons.

37 For an excellent account of this distinguished man, see Youmans's Introduction to the Correlation and Conservation of Forces, p. xvii.

38 DRAPER, J. W., *loc. cit.*   •

39 HENRY, JOSEPH, Agric. Rep. Patent Office, 1857, 440.

40 WATTERS, J. H., An Essay on Organic, or Life-force. Written for the degree of Doctor of Medicine in the University of Pennsylvania, Philadelphia, 1851. See also St. Louis Medical and Surgical Journal, II, v, Nos. 3 and 4, 1868; Dec. 1868, and Nov. 10, 1869.

41 LeCONTE, JOSEPH, The Correlation of Physical, Chemical and Vital Force, and the Conservation of Force in Vital Phenomena. American Journal of Science, II, xxviii, 305, Nov. 1859.

42 LOMBARD, J. S., *loc. cit.*

43 NOYES, T. R., *loc. cit.*

44 WOOD, L. H., *loc. cit.*

AS REGARDS PROTOPLASM, ETC.

PREFATORY NOTE.

The substance of the greater part of this paper, which has been in the present form for some time, was delivered, as a lecture, at a Conversazione of the Royal College of Physicians of Edinburgh, in the Hall of the College, on the evening of Friday, the 30th of April last.

It will be found to support itself, so far as the facts are concerned, on the most recent German physiological literature, as represented by Rindfleisch, Kühne, and especially Stricker, with which last, for the production of his "Handbuch," there is associated every great histological name in Germany.

EDINBURGH, *October*, 1869.

LIBRARY UNIVERSITY C... CALIFORNIA

As Regards Protoplasm, etc.

It is a pleasure to perceive Mr. Huxley open his clear little essay with what we may hold, perhaps, to be the manly and orthodox view of the character and products of the French writer, Auguste Comte. "In applying the name of 'the new philosophy' to that estimate of the limits of philosophical inquiry which he" (Professor Huxley), " in common with many other men of science, holds to be just," the Archbishop of York confounds, it seems, this new philosophy with the Positive philosophy of M. Comte ; and thereat Mr. Huxley expresses himself as greatly astonished. Some of us, for our parts, may be inclined at first to feel astonished at Mr. Huxley's astonishment ; for the school to which, at least on the philosophical side, Mr. Huxley seems to belong, is even notorious for its postration before Auguste Comte, whom, especially, so far as method and systematization are concerned, it regards as the greatest intellect since Bacon. For such, as it was the opinion of Mr. Buckle, is understood to be the opinion also of Messrs. Grote, Bain, and Mill. In fact, we may say that such is commonly and currently considered the characteristic and distinctive opinion of that whole perverted or inverted reaction which has been called the *Revulsion*. That is to say, to give this word a moment's explanation, that the Voltaires and Humes and Gibbons having long enjoyed an immunity of sneer at man's blind pride and

wretched superstition—at *his* silly non-natural honor
and *her* silly non-natural virtue—a reaction had set in,
exulting in poetry, in the splendor of nature, the nobleness of man, and the purity of woman, from which reaction again we have, almost within the last decennium,
been revulsively, as it were, called back,—shall we say
by some "bolder" spirits—the Buckles, the Mills, &c.?
—to the old illumination or enlightenment of a hundred
years ago, in regard to the weakness and stupidity of
man's pretensions over the animality and materiality
that limit him.　Of this revulsion, then, as said, a main
feature, especially in England, has been prostration
before the vast bulk of Comte ; and so it was that Mr.
Huxley's protest in this reference, considering the philosophy he professed, had that in it to surprise at first.
But if there was surprise, there was also pleasure ; for
Mr. Huxley's estimate of Comte is undoubtedly the
right one.　"So far as I am concerned," he says, "the
most reverend prelate" (the Archbishop of York)
"might dialectically hew M. Comte in pieces as a modern Agag, and I should not attempt to stay his hand ;
for, so far as my study of what specially characterizes
the Positive philosophy has led me, I find therein little
or nothing of any scientific value, and a great deal
which is as thoroughly antagonistic to the very essence
of science as anything in ultramontane Catholicism."
"It was enough," he says again, "to make David Hume
turn in his grave, that here, almost within earshot of
his house, an instructed audience should have listened
without a murmur while his most characteristic doctrines were attributed to a French writer of fifty years'
later date, in whose dreary and verbose pages we miss
alike the vigor of thought and the exquisite clearness

of style of the man whom I make bold to term the
most acute thinker of the eighteenth century—even
though that century produced Kant."

Of the doctrines themselves which are alluded to
here, I shall say nothing now; but of much else that is
said, there is only to be expressed a hearty and even
gratified approval. I demur, to be sure, to the exalta-
tion of Hume over Kant—high as I place the former.
Hume, with infinite fertility, surprised us, it may be
said, perhaps, into attention on a great variety of points
which had hitherto passed unquestioned; but, even on
these points, his success was of an interrupted, scattered
and inconclusive nature. He set the world adrift, but
he set man too, reeling and miserable, adrift with it.
Kant, again, with gravity and reverence, desired to refix,
but in purity and truth, all those relations and institu-
tions which alone give value to existence—which alone
are humanity, in fact—but which Hume, with levity and
mockery, had approached to shake. Kant built up
again an entire new world for us of knowledge and
duty, and, in a certain way, even belief; whereas Hume
had sought to dispossess us of every support that man
as man could hope to cling to. In a word, with *at least*
equal fertility, Kant was, as compared with Hume, a
graver, deeper, and, so to speak, a more consecutive,
more comprehensive spirit. Graces there were indeed,
or even, it may be said, subtleties, in which Hume had
the advantage perhaps. He is still in England an
unsurpassed master of expression—this, certainly, in
his History, if in his Essays he somewhat baffles his
own self by a certain labored breadth of conscious fine
writing, often singularly inexact and infelicitous. Still
Kant, with reference to his products, must be allowed

much the greater importance. In the history of philosophy he will probably always command as influential a place in the modern world as Socrates in the ancient ; while, as probably, Hume will occupy at best some such position as that of Heraclitus or Protagoras. Hume, nevertheless, if equal to Kant, must, in view at once of his own subjective ability and his enormous influence, be pronounced one of the most important of writers. It would be difficult to rate too high the value of his French predecessors and cotemporaries as regards purification of their oppressed and corrupt country; and Hume must be allowed, though with less call, to have subserved some such function in the land we live in. In preferring Kant, indeed, I must be acquitted of an undue partiality; for all that appertains to personal bias was naturally, and by reason of early and numerous associations, on the side of my countryman.

Demurring, then, to Mr. Huxley's opinion on this matter, and postponing remark on the doctrines to which he alludes, I must express a hearty concurrence with every word he utters on Comte. In him I too "find little or nothing of any scientific value." I too have been lost in the mere mirage and sands of "those dreary and verbose pages ;" and I acknowledge in Mr. Huxley's every word the ring of a genuine experience. M. Comte was certainly a man of some mathematical and scientific proficiency, as well as of quick but biased intelligence. A member of the *Aufklärung*, he had seen the immense advance of physical science since Newton, under, as is usually said, the method of Bacon ; and, like Hume, like Reid, like Kant, *who had all anticipated him in this*, he sought to transfer that method to the domain of mind. In this he failed ; and though in a sociological aspect he is not without true glances into the present disintegration of society and the conditions of it, anything of importance cannot be claimed for him. There is not a sentence in his book that, in the hollow elaboration and windy pretentiousness of its build, is not an exàct type of its own constructor. On the whole, indeed, when we consider the little to which he attained, the empty inflation of his claims, the monstrous and maniacal self-conceit into which he was *exalted*, it may appear, perhaps, that charity to M. Comte himself, to say nothing of the world, should induce us to wish that both his name and his works were buried in oblivion. Now, truly, that Mr. Huxley (the " call" being for the moment his) has so pronounced himself, especially as the facts of the case are exactly and absolutely what he indicates, perhaps we may expect this consummation not to be so very long delayed. More than those members of the revulsion already mentioned, one is apt to suspect, will be anxious now to beat a retreat. Not that this, however, is so certain to be allowed them ; for their estimate of M. Comte is a valuable element in the estimate of themselves.

·Frankness on the part of Mr. Huxley is not limited to his opinion of M. Comte ; it accompanies us throughout his whole essay. He seems even to take pride, indeed, in naming always and everywhere his object at the plainest. That object, in a general point of view, relates, he tells us, solely to materialism, but with a double issue. While it is his declared purpose, in the first place, namely, to lead us into materialism, it is equally his declared purpose, in the second place, to lead us out of materialism. On the first issue, for

example, he directly warns his audience that to accept
the conclusions which he conceives himself to have
established on Protoplasm, is to accept these also:
That "all vital action" is but "the result of the molecular forces" of the physical basis; and that, by consequence, to use his own words to his audience, "the
thoughts to which I am now giving utterance, and your
thoughts regarding them, are but the expression of
molecular changes in that matter of life which is the
source of our other vital phenomena." And, so far,
I think, we shall not disagree with Mr. Huxley when
he says that "most undoubtedly the terms of his propositions are distinctly materialistic." Still, on the second
issue, Mr. Huxley asserts that he is "individually no
materialist." "On the contrary, he believes materialism to involve grave philosophical error;" and the
"union of materialistic terminology with the repudiation of materialistic philosophy" he conceives himself
to share "with some of the most thoughtful men with
whom he is acquainted." In short, to unite both issues,
we have it in Mr. Huxley's own words, that it is the
single object of his essay "to explain how such a union
is not only consistent with, but necessitated by, sound
logic;" and that, accordingly, he will, in the first place,
"lead us through the territory of vital phenomena to
the materialistic slough," while pointing out, in the second, "the sole path by which, in his judgment, extrication is possible." Mr. Huxley's essay, then, falls evidently into two parts; and of these two parts we may
say, further, that while the one—that in which he leads
us into materialism—will be predominatingly physiological, the other—or that in which he leads us out
of materialism—will be predominatingly philosophical.

Two corresponding parts would thus seem to be pre-
scribed to any full discussion of the essay; and of
these, in the present needs of the world, it is evidently
the latter that has the more promising theme. The
truth is, however, that Mr. Huxley, after having exerted
all his strength in his first part to throw us into "the
materialistic slough," by *clear necessity of knowledge*,
only calls to us, in his second part, to come out of this
slough again, on the somewhat *obscure necessity of ignorance*. This, then, is but a lop-sided balance, where a
scale in the air only seems to struggle vainly to raise
its well-weighted fellow on the ground. Mr. Huxley,
in fact, possesses no remedy for materialism but what
lies in the expression that, while he knows not what
matter is in itself, he certainly knows that casualty is
but contingent succession; and thus, like the so-called
"philosophy" of the Revulsion, Mr. Huxley would only
mock us into the intensest dogmatism on the one side
by a fallacious reference to the intensest scepticism on
the other.

The present paper, then, will regard mainly Mr. Huxley's argument *for* materialism, but say what is required,
at the same time, on his alleged argument—which is
merely the imaginary, or imaginative, impregnation of
ignorance—*against* it.

Following Mr. Huxley's own steps in his essay, the
course of his positions will be found to run, in summary, thus :—

What is meant by the physical basis of life is, that
there is one kind of matter common to all living beings,
and it is named protoplasm. No doubt it may appear
at first sight that, in the various kinds of living beings,
we have only *difference* before us, as in the lichen on the

rock and the painter that paints it,—the microscopic animalcule or fungus and the Finner whale or Indian fig,—the flower in the hair of a girl and the blood in her veins, etc. Nevertheless, throughout these and all other diversities, there really exists a threefold *unity*—a unity of faculty, a unity of form, and a unity of substance.

On the first head, for example, or as regards faculty, power, the action exhibited, there are but three categories of *human* activity—contractility, alimentation, and reproduction ; and there are no fewer for the *lower* forms of life, whether animal or vegetable. In the nettle, for instance, we find the woody case of its sting lined by a granulated, semi-fluid layer, that is possessed of contractility. But in this respect—that is, in the possession of contractile substance—other plants are as the nettle, and all animals are as plants. Protoplasm—for the nettle-layer alluded to is protoplasm—is common to the whole of them. The difference, in short between the powers of the lowest plant or animal and those of the highest is one only of degree and not of kind.

But, on the second head, it is not otherwise in form, or manifested external appearance and structure. Not the sting only, but the whole nettle, is made up of protoplasm ; and of all the other vegetables the nettle is but a type. Nor are animals different. The colorless blood-corpuscles in man and the rest are identical with the protoplasm of the nettle ; and both he and they consisted at first only of an aggregation of such. Protoplasm is the common constituent—the common origin. At last, as at first, all that lives, and every part of all that lives, are but nucleated or unnucleated, modified or unmodified, protoplasm.

But, on the third head, or with reference to unity of

substance, to internal composition, chemistry establishes this also. All forms of protoplasm, that is, consist alike of carbon, hydrogen, oxygen, and nitrogen, and behave similarly under similar reagents.

So, now, a uniform character having in this threefold manner been proved for protoplasm, what is its origin, and what its fate? Of these the latter is not far to seek. The fate of protoplasm is death—death into its chemical constituents; and this determines its origin also. Protoplasm can originate only in that into which it dies,—the elements—the carbon, hydrogen, oxygen, and nitrogen—of which it was found to consist. Hydrogen, with oxygen, forms water; carbon, with oxygen, carbonic acid; and hydrogen, with nitrogen, ammonia. Similarly, water, carbonic acid and ammonia form, in union, protoplasm. The influence of pre-existing protoplasm only determines combination in *its* case, as that of the electric spark determines combination in the case of water. Protoplasm, then, is but an aggregate of physical materials, exhibiting in combination—only as was to be expected—new properties. The properties of water are not more different from those of hydrogen and oxygen than the properties of protoplasm are different from those of water, carbonic acid, and ammonia. We have the same warrant to attribute the consequences to the premises in the one case as in the other. If, on the first stage of combination, repre-sented by that of water, *simples* could unite into some-thing so different from themselves, why, on the second stage of combination, represented by that of proto-plasm, should not *compounds* similarly unite into some-thing equally different from themselves? If the con-stituents are credited with the properties *there*, why refuse to credit the constituents with the properties *here?* To the constituents of protoplasm, in truth, any new element, named vitality, has no more been added, than to the constituents of water any new element, named aquosity. Nor is there any logical halting-place between this conclusion and the further and final one: That all vital action whatever, intellectual included, is but the result of the molecular forces of the protoplasm which displays it.

These sentences will be acknowledged, I think, fairly to represent Mr. Huxley's relative deliverances, and, consequently, as I may be allowed to explain again, the only important—while much the larger—part of the whole essay. Mr. Huxley, that is, while devoting fifty paragraphs to our physiological immersion in the " materialistic slough," grants but one-and twenty towards our philosophical escape from it; the fifty besides being, so to speak, in reality the wind, and the one-and-twenty only the whistle for it. What these latter say, in effect, is no more than this, that,—matter being known not in itself but only in its qualities, and cause and effect not in their nexus but only in their sequence,—matter may be spirit or spirit matter, cause effect or effect cause—in short, for aught that Mr. Huxley more than phenomenally knows, this may be that or that this, first second, or second first, but the conclusion shall be this, that he will lay out all our knowledge materially, and we may lay out all our ignorance immaterially—if we will. Which reasoning and conclusion, I may merely remark, come precisely to this: That Mr. Huxley—who, hoping yet to see each object (a pin, say) not in its qualities but in *itself,* still, consistently antithetic, cannot believe in the extinction of fire by water or of life by the rope, for any *reason* or for any *necessity* that lies in the nature of the case, but simply for the habit of the thing—has not yet put himself at home with the metaphysical categories of *substance* and *casualty;* thanks, perhaps, to those guides of ‚his whom we, the amusing Britons that we are, bravely proclaim "the foremost thinkers of the day" !

The matter and manner of the whole essay are now fairly before us, and I think that, with the approbation of the reader, its procedure, generally, may be described as an attempt to establish, not by any complete and systematic induction, but by a variety of partial and illustrative assertions, two propositions. Of these propositions the first is, That all animal and vegetable organisms are essentially alike in power, in form, and in substance ; and the second, That all vital and intellectual functions are the properties of the molecular disposition and changes of the material basis (protoplasm) of which the various animals and vegetables consist. In both propositions, the agent of proof is this same alleged material basis of life, or protoplasm. For the first of them, all animal and vegetable organisms shall be identified in protoplasm ; and for the second, a simple chemical analogy shall assign intellect and vitality to the molecular constituents of the protoplasm, in connection with which they are at least exhibited.

In order, then, to obtain a footing on the ground offered us, the first question we naturally put is, What is Protoplasm ? And an answer to this question can be obtained only by a reference to the historical progress of the physiological cell theory.

That theory may be said to have wholly grown up

since John Hunter wrote his celebrated work 'On the
Nature of the Blood,' etc. New growths, to Hunter,
depended on an exudation of the plasma of the blood,
in which, by virtue of its own *plasticity*, vessels formed,
and conditioned the further progress. The influence of
these ideas seems to have still acted, even after a conception of the cell was arrived at. For starting element,
Schleiden required an intracellular plasma, and Schwann
a structureless exudation, in which minute granules, if
not indeed already pre-existent, formed, and by aggregation grew into nuclei, round which singly the production of a membrane at length enclosed a cell. It was
then that, in this connection, we heard of the terms
blastema and cyto-blastema. The theory of the vegetable cell was completed earlier than that of the animal
one. Completion of this latter, again, seems to have
been first effected by Schwann, after Müller had insisted
on the analogy between animal and vegetable tissue,
and Valentin had demonstrated a nucleus in the animal
cell, as previously Brown in the vegetable one. But
assuming Schwann's labor, and what surrounded it, to
have been a first stage, the wonderful ability of Virchow
may be said to have raised the theory of the cell fully
to a second stage. Now, of this second stage, it is the
dissolution or resolution that has led to the emergence
of the word Protoplasm.

The body, to Virchow, constituted a free state of individual subjects, with equal rights but unequal capacities. These were the cells, which consisted each of
an enclosing membrane, and an enclosed nucleus with
surrounding intracellular matrix or matter. These
cells, further, propagated themselves, chiefly by partition
or division ; and the fundamental principle of the whole

theory was expressed in the dictum, " *Omnis cellula e cellulâ.*" That is, the nucleus, becoming gradually elongated, at last parted in the midst ; and each half, acting as center of attraction to the surrounding intracellular matrix or contained matter, stood forth as a new nucleus to a new cell, formed by division at length of the original cell.

The first step taken in resolution of this theory was completed by Max Schultze, preceded by Leydig. This was the elimination of an investing membrane. Such membrane may, and does, ultimately form ; but in the first instance, it appears, the cell is naked. The second step in the resolution belongs perhaps to Brücke, though preceded by Bergmann, and though Max Schultze, Kühne, Hæckel, and others ought to be mentioned in the same connection. This step was the elimination, or at least subordination, of the nucleus. The nucleus, we are to understand now, is necessary neither to the division nor to the existence of the cell.

Thus, then, stripped of its membrane, relieved of its nucleus, what now remains for the cell? Why, nothing but what *was* the contained matter, the intracellular matrix, and *is*—Protoplasm.

In the application of this word itself, however, to the element in question, there are also a step or two to be noticed. The first step was Dujardin's discovery of sarcode ; and the second the introduction of the term protoplasm as the name for the layer of the *vegetable* cell that lined the cellulose, and enclosed the nucleus. Sarcode, found in certain of the lower forms of life, was a simple substance that exhibited powers of spontaneous contraction and movement. Thus, processes of such simple, soft, contractile matter are protruded by the

rhizopods, and locomotion by their means effected.
Remak first extended the use of the term protoplasm
from the layer which bore that name in the vegetable
cell to the analogous element in the animal cell ; but it
was Max Schultze, in particular, who, by applying the
name to the intracellular matrix, or contained matter,
when divested of membrane, and by identifying this
substance itself with sarcode, first fairly established protoplasm, name and thing, in its present prominence.

In this account I have necessarily omitted many subordinate and intervening steps in the successive estab-
lishment of the *contractility*, superior *importance*, and
complete *isolation* of this thing to which, under the
name of protoplasm, Mr. Huxley of late has called
such vast attention. Besides the names mentioned,
there are others of great eminence in this connection,
such as Meyen, Siebold, Reichert, Ecker, Henle, and
Kölliker among the Germans ; and among ourselves,
Beale and Huxley himself. John Goodsir will be mentioned again.

We have now, perhaps, obtained a general idea of
protoplasm. Brücke, when he talks of it as "living
cell-body or elementary organism," comes very near the
leading idea of Mr. Huxley as expressed in his phrase,
"the physiological basis, or matter, of life." Living
cell-body, elementary organism, primitive living matter
—that, evidently, is the quest of Mr. Huxley. There is
aqueous matter, he would say, perhaps, composed of
hydrogen and oxygen, and it is the same thing whether
in the rain-drop or the ocean ; so, similarly, there is
vital matter, which, composed of carbon, hydrogen, oxygen, and nitrogen, is the same thing whether in crypto-
gams or in elephants, in animalcules or in men. What,

in fact, Mr. Huxley seeks, probably, is living protein—
protein, so to speak, struck into life.   Just such appears
to him to be the nature of protoplasm, and in it he believes himself to possess at last *a living clay* wherewith
to build the whole organic world.

The question, What is Protoplasm? is answered,
then ; but, for the understanding of what is to follow,
there is still one general consideration to be premised.

Mr. Huxley's conception of protoplasm, as we have
seen, is that of living matter, living protein ; what we
may call, perhaps, elementary life-stuff.   Now, is it
quite certain that Mr. Huxley is correct in this conception?   Are we to understand, for example, that cells
have now definitively vanished, and left in their place
only a uniform and universal *matter* of quite indefinite
proportions?   No ; such an understanding would be
quite wrong.   Whatever may be the opinion of the ad-
herents of the molecular theory of generation, it is certain that all the great German histologists still hold by
the cell, and can hardly open their mouths without men-
tion of it.   I do not allude here to any special adhe-
rents of either nucleus or membrane, but to the most
advanced innovators in both respects ; to such men as
Schultze and Brücke and Kühne.   These, as we have
seen, pretty well confine their attention, like Mr. Huxley, to the protoplasm.   But they do not the less on
that account talk of the cell.   For them, it is only in
cells that protoplasm exists.   To their view, we cannot
fancy protoplasm as so much matter in a pot, in an oint-
ment-box, any portion of which scooped out in an ear-
picker would be so much life-stuff, and, though a part,
quite as good as the whole.   This seems to be Mr.
Huxley's conception, but it is not theirs.   A certain

measure goes with protoplasm to constitute it an organism to them, and worthy of their attention. They refuse to give consideration to any mere protoplasm-*shred* that may not have yet ceased, perhaps, to exhibit all sign of contractility under the microscope, and demand a protoplasm-*cell.* In short, protoplasm is to them still distributed into cells, and only that measure of protoplasm is cell that is adequate to the whole group of vital manifestations. Brücke, for example, of all innovators probably the most innovating, and denying, or inclined to deny, both nucleus and membrane, does not hesitate, according to Stricker, to speak still of cells as self-complete organisms, that move and grow, that nourish and reproduce themselves, and that perform specific function. "Omnis cellula e cellulâ," is the rubric they work under as much now as ever. The heart of a turtle, they say, is not a turtle ; so neither is a protoplasm-shred a protoplasm-cell.

This, then, is the general consideration which I think it necessary to premise ; and it seems, almost of itself, to negate Mr. Huxley's reasonings in advance, for it warrants us in denying that physiological clay of which all living things are but bricks baked, Mr. Huxley intimates, and in establishing in its place cells as before— living cells that differ infinitely the one from the other, and so differ from the very first moment of their existence. This consideration shall not be allowed to pretermit, however, an examination of Mr. Huxley's own proofs, which will only the more and more avail to indicate the difference suggested.

These proofs, as has been said, would, by means of the single fulcrum of protoplasm, establish, first, the identity, and, second, the materiality, of all vegetable

and animal life. These are, shortly, the two propositions which we have already seen, and to which, in their
order, we now pass.

All organisms, then, whether animal or vegetable,
have been understood for some time back to originate
in and consist of cells ; but the progress of physiology
has *secmed* now to substitute for cells a single matter of
life, protoplasm ; and it is here that Mr. Huxley sees his
cue.   Mr. Huxley's very first word is the "physical basis
or matter of life ;" and he supposes "that to many the
idea that there is such a thing may be novel."   This, then,
so far, is what is *new* in Mr. Huxley's contribution.   He
seems to have said to himself, if formerly the whole
world was thought kin in an "ideal" or formal element,
organization, I shall now finally complete this identification in a "physical" or material element, protoplasm.
In short, what at this stage we are asked to witness in
the essay is, the identification of all living beings whatever in the identity of protoplasm.   As there is a
single matter, clay, which is the matter of all bricks, so
there is a single matter, protoplasm, which is the matter
of all organisms.   "Protoplasm is the clay of the potter, which, bake it and paint it as he will, remains clay,
separated by artifice, and not by nature, from the commonest brick or sun-dried clod."   Now here I cannot
help stopping a moment to remark that Mr. Huxley
puts emphatically his whole soul into this sentence, and
evidently believes it to be, if we may use the word, a
clincher.   But, after all, does it say much? or rather,
does it say anything?   To the question, "Of what are
you made?" the answer, for a long time now, and by
the great mass of human beings who are supposed civilized, has been "Dust."   Dust, and the same dust, has

been allowed to constitute us all. But materialism has not on that account been the irresistible result. Attention hitherto—and surely excusably, or even laudably in such a case—has been given not so much to the dust as to the "potter," and the "artifice" by which he could so transform, or, as Mr. Huxley will have it, *modify* it. To ask us to say, instead of dust, clay, or even protoplasm, is not to ask us for much, then, seeing that even to Mr. Huxley there still remain both the "potter" and his "artifice."

But to return: To Mr. Huxley, when he says all bricks, being made of clay, are the same thing, we answer, Yes, undoubtedly, if they are made of the same clay. That is, the bricks are identical if the clay is identical; but, on the other hand, by as much as the clay differs will the bricks differ. And, similarly, all organisms can be identified only if their composing protoplasm can be identified. To this stake is the argument of Mr. Huxley bound.

This argument itself takes, as we have seen, a threefold course : Mr. Huxley will prove his position in this place by reference, firstly, to unity of faculty ; secondly, to unity of form ; and thirdly, to unity of substance. It is this course of proof, then, which we have now to follow, but taking the question of substance, as simplest, first, and the others later.

By substance, Mr. Huxley understands the internal or chemical composition ; and, with a mere reference to the action of reagents, he asserts the protoplasm of all living beings to be an identical combination of carbon, hydrogen, oxygen, and nitrogen. It is for us to ask, then, Are all samples of protoplasm identical, first, in their chemical composition, and, second, under the action of the various reagents?

On the first clause, we may say, in the first place, towards a proof of difference which will only cumulate, I hope, that, even should we grant in all protoplasm an identity of chemical ingredients, what is called *Allotropy* may still have introduced no inconsiderable variety. Ozone is not antozone, nor is oxygen either, though in chemical constitution all are alike. In the second place, again, we may say that, with *varying proportions*, the same component parts produce very various results. By way of illustration, it will suffice to refer to such different things as the proteids, gluten, albumen, fibrin, gelatine, etc., compared with the urinary products, urea and uric acid ; or with the biliary products, glycocol, glycocolic acid, bili-rubin, bili-verdin, etc. ; and yet all these substances, varying so much the one from the other, are, as protoplasm is, compounds of carbon, hydrogen, oxygen, and nitrogen. But, in the third place, we are not limited to a *may say ;* we can assert the fact that all protoplasm is not chemically identical. All the tissues of the organism are called protoplasm by Mr. Huxley ; but can we predicate chemical identity of muscle and bone, for example? In such cases Mr. Huxley, it is true, may bring the word "modified" into use ; but the objection of modification we shall examine later. In the mean time, we are justified, by Mr. Huxley's very argument, in regarding all organized tissues whatever as protoplasm ; for if these tissues are not to be identified in protoplasm, we must suppose denied what it was his one business to affirm. And it is against that affirmation that we point to the fact of much chemical difference obtaining among the tissues, not only in the *proportions* of their fundamental elements, but also in the *addition* (and proportions as well) of such others as chlorine, sulphur, phosphorus, potass, soda, lime, magnesia, iron, etc. Vast differences vitally must be legitimately assumed for tissues that are so different chemically. But, in the fourth place, we have the authority of the Germans for asserting that the cells themselves—and they now, to the most advanced, are only protoplasm — do differ chemically, some being found to contain glycogen, some cholesterine, some protogon, and some myosin. Now such substances, let the chemical analogy be what it may, must still be allowed to introduce chemical difference. In the last place, Mr. Huxley's analysis is an analysis of *dead* protoplasm, and indecisive, consequently, for that which lives. Mr. Huxley betrays sensitiveness in advance to this objection ; for he seeks to rise above the sensitiveness and the objection at once by styling the latter "frivolous." Nevertheless the Germans say pointedly that it is unknown whether the same elements are to be referred to the cells after as before death. Kühne does not consider it proved that living muscle contains syntonin ; yet Mr. Huxley tells us, in his Physiology, that "syntonin is the chief constituent of muscle and flesh." In general, we may say, according to Stricker, that all weight is put now on the examination of living tissue, and that the difference is fully allowed between that and dead tissue.

On the second clause now, or with regard to the action of reagents, these must be denied to produce the like result on the various forms of protoplasm. With reference to temperature, for example, Kühne reports the movements of the amoeba to be arrested in iced water ; while, in the same medium, the ova of the trout furrow famously, but perish even in a warmed room. Others, again, we are told, may be actually dried, and yet live. Of ova in general, in this connection, it is said that they live or die according as the temperature to which they are exposed differs little or much from that which is natural to the organisms producing them. In some, according to Max Schultze, even distilled water is enough to arrest movement. Now, not to dwell longer here, both amoeba and ova are to Mr. Huxley pure protoplasm; and such difference of result, according to difference of temperature, etc., must assuredly be allowed to point to a difference of original nature. Any conclusion so far, then, in regard to unity of substance, whether the chemical composition or the action of reagents be considered, cannot be said to bear out the views of Mr. Huxley.

What now of the unities of form and power in protoplasm? By form, Mr. Huxley will be found to mean the general appearance and structure ; and by faculty or power, the action exhibited. Now it will be very easy to prove that, in neither respect, do all specimens of protoplasm agree. Mr. Huxley's representative protoplasm, it appears, is that of the nettle-sting ; and he describes it as a granulated, semi-fluid body, contractile in mass, and contractile also in detail to the development of a species of circulation. Stricker, again, speaks of it as a homogeneous substance, in which any granules that may appear must be considered of foreign importation, and in which there are no evidences of circulation. In this last respect, then, that Mr. Huxley should talk of "tiny Maelstroms," such as even in the silence of a tropical noon might stun us, if heard, as "with the roar of a great city," may be viewed, perhaps, as a rise into poetry beyond the occasion.

Further, according to Stricker, protoplasm varies al-.

most infinitely in consistence, in shape, in structure, and in function. In consistence, it is sometimes so fluid as to be capable of forming in drops ; sometimes semi-fluid and gelatinous ; sometimes of considerable resist-ance. In shape—for to Stricker the cells are now pro-toplasm—we have club-shaped protoplasm, globe-shaped protoplasm, cup-shaped protoplasm, bottle-shaped proto-plasm, spindle-shaped protoplasm—branched, threaded, ciliated protoplasm, — circle-headed protoplasm — flat, conical, cylindrical, longitudinal, prismatic, polyhedral, and palisade-like protoplasm. In structure, again, it is sometimes uniform and sometimes reticulated into inter-spaces that contain fluid. In function, lastly—and here we have entered on the consideration of faculty or power —some protoplasm is vagrant (so to translate *wan-dernd)*, and of unknown use, like the colorless blood-corpuscles.

In reference to these, as strengthening the argument, and throwing much light generally, I break off a mo-ment to say that, very interesting as they are in them-selves, and as Recklinghausen, in especial, has made them, Mr. Huxley's theory of them disagrees consider-ably with the prevalent German one. He speaks of them as the source of the body in general, yet, in his Physiology, he talks of the spleen, the lymphatics, and even the liver—*parts* of the body—as *their* source. They are so few in number that, while Mr. Huxley is thankful to be able to point to the inside of the lips as a seat for them, they bear to the red corpuscles only the proportion of 1 to 450. This disproportion, how-ever, is no bar to Mr. Huxley's derivation of the latter from the former. But the fact is questioned. The Germans, generally, for their part, describe the colorless, or vagrant, blood-corpuscles as probably media of conjugation or reparation, but acknowledge their function to be as yet quite unknown ; while Rindfleisch, characterizing the spleen as the grave of the red, and the womb of the white, corpuscles, evidently refers the latter to the former. This, indeed, is a matter of direct assertion with Preyer, who has " shown that pieces of red blood-corpuscles may be eaten by the amoeboid cells of the frog," and holds that the latter (the white corpuscles) proceed directly from the former (the red corpuscles) ; so that it seems to be determined in the mean time that there is no proof of the reverse being the fact.

In function, then, to resume, some protoplasm is vagrant, and of unknown use. Some again produces pepsine, and some fat. Some at least contains pigment. Then there is nerve-protoplasm, brain-protoplasm, bone-protoplasm, muscle-protoplasm, and protoplasm of all the other tissues, no one of which but produces only its own kind, and is uninterchangeable with the rest. Lastly, on this head, we have to point to the overwhelming fact that there is the infinitely different protoplasm of the various infinitely different plants and animals, in each of which its own protoplasm, as in the case of that of the various tissues, but produces its own kind, and is uninterchangeable with that of the rest.

It may be objected, indeed, that these latter are examples of modified profoplasm. The objection of modification, as said, we have to see by itself later ; but, in the mean time, it may be asked, Where are we to begin, *not* to have modified protoplasm? We have the example of Mr. Huxley himself, who, in the nettle-

sting, begins already with modified protoplasm ; and
we have the authority of Rindfleisch for asserting that
" in every different tissue we must look for a different
initial term of the productive series." This, evidently,
is a very strong light on the original multiplicity of
protoplasm, which the consideration, as we have seen,
of the various plants and animals, has made, further,
infinite. This is enough ; but there is no wish to evade
beginning with the very beginning—with absolutely
pure initial protoplasm, if it can but be given us in any
reference. The simple egg—that, probably is the beginning—that, probably, is the original identity ; yet
even there we find already distribution of the identity
into infinite difference. This, certainly, with reference
to the various organisms, but with reference also to the
various tissues. That we regard the egg as the beginning, and that we do not start, like the smaller exceptional physiological school, with molecules themselves,
depends on this, that the great Germans so often alluded to, Kühne among them, still trust in the experiments of Pasteur ; and while they do not deny the possibility, or even the fact, of molecular generation, still
feel justified in denying the existence of any observation that yet unassailably attests a *generatio æquivoca.*
By such authority as this the simple philosophical spec-
tator has no choice but to take his stand ; and therefore
it is that I assume the egg as the established beginning,
so far, of all vegetable and animal organisms. To the
egg, too, as the beginning, Mr. Huxley, though the
lining of the nettle-sting is his representative protoplasm, at least refers. " In the earliest condition of
the human organism," he says, in allusion to the white
(vagrant) corpuscles of the blood, " in that state in

which it has but just become distinguished from the
egg in which it arises, it is nothing but an aggregation
of such corpuscles, and every organ of the body was
once no more than such an aggregation." Now, in beginning with the egg—an absolute beginning being denied us in consequence of the pre-existent infinite
difference of the egg or eggs themselves—we may
gather from the German physiologists some such account of the actual facts as this.

The first change signalized in the impregnated egg
seems that of *Furchung*, or furrowing—what the Germans call the *Furchungskugeln*, the *Dotterkugeln*, form.
Then these *Kugeln*—clumps, eminences, monticles, we
may translate the word—break into cells ; and these
are the cells of the embryo. Mr. Huxley, as quoted,
refers to the whole body, and every organ of the body,
as at first but an aggregation of colorless blood-corpuscles ; but in the very statement which would render
the identity alone explicit, the difference is quite as
plainly implicit. As much as this lies in the word " organs," to say nothing of "human." The cells of the
"organs," to which he refers, are even then uninterchangeable, and produce but themselves. The Germans tell us of the *Keimblatt*, the germ-leaf, in which
all these organs originate. This *Blatt*, or leaf, is threefold, it seems ; but even these folds are not indifferent.
The various cells have their distinct places in them from ·
the first. While what in this connection are called the
epithelial and endorthelial tissues spring respectively
from the *upper* and *under* leaf, connective tissues, with
muscle and blood, spring from the *middle* one. Surely
in such facts we have a perfect warrant to assert the
initial non-identity of protoplasm, and to insist on this,

that, from the very earliest moment—even literally *ab ovo*—brain-cells only generate brain-cells, bone-cells bone-cells, and so on.

These considerations on function all concern faculty or power ; but we have to notice now that the characteristic and fundamental form of power is to Mr. Huxley *contractility.* He even quotes Goethe in proof of contractility being the main power or faculty of *Man !* Nevertheless it is to be said at once that, while there are differences in what protoplasm *is* contractile, all protoplasm is not contractile, nor dependent on contractility for its functions. In the former respect, for example, muscle, while it is the contractile tissue special, is also to Mr. Huxley protoplasm ; yet Stricker asserts the inner construction of the contractile substance, of which muscle-fibre virtually consists, to be essentially different from contractile protoplasm. Here, then, we have the contractile *substance* proper "essentially different" from the contractile *source* proper. In the latter respect, again, we shall not call in the *un*contractible substances which Mr. Huxley. himself denominates protoplasm—bread, namely, roast mutton, and boiled lobster ; but we may ask where—even in the case of a living body—is the contractility of white of egg ? In this reference, too, we may remark that Kühne, who divides the protoplasm of the epidermis into three classes, has been unable to distinguish contractility in his own third class. Lastly, where, in relation to the protoplasm of the nervous system, is there evidence of its contractility ? Has any one pretended that thought is but the contraction of the brain ; or is it by contraction that the very nerves operate contraction—the nerves that supply muscles, namely ? Mr. Huxley himself, in his Physiology, describes nervous action very differently. There *conduction* is spoken of without a hint of contraction. Of the higher faculties of man I have to speak again; but let us just ask where, in the case of any pure sensation—smell, taste, touch, sound, color—is there proof of any contraction? Are we to suppose that between the physical cause of heat without and the mental sensation of heat within, contraction is anywhere interpolated? Generally, in conclusion here, while reminding of Virchow's testimony to the inherent inequalities of cell-capacity, let us but, on the question of faculty, contrast the kidney and the brain, even as these organs are viewed by Mr. Huxley. To him the one is but a sieve for the extrusion of refuse : the other thinks Newton's 'Principia' and Iliads of Homer.

Probably, then, in regard to any continuity in protoplasm of power, of form, or of substance, we have seen *lacunae* enow. Nay, Mr. Huxley himself can be adduced in evidence on the same side. Not rarely do we find in his essay admissions of *probability* where it is *certainty* that is alone in place. He says, for example, " It is more than probable that *when* the vegetable world *is* thoroughly explored we *shall* find all plants in possession of the same powers." When a conclusion is decidedly announced, it is rather disappointing to be told, as here, that the premises are still to collect. *" So far,"* he says again, " as the conditions of the manifestations of the phenomena of contractility have *yet* been studied." Now, such a *so far* need not be *very far ;* and we may confess in passing, that from Mr. Huxley the phrase, " the conditions of the *manifestations* of the *phenomena*," grates. We hear again that it is " the rule *rather* than the exception," or that " weighty authorities have *suggested"* that such and such things "probably occur," or, while contemplating the nettle-sting, that such "*possible* complexity" in other cases "*dawns* upon one." On other occasions he expresses himself to the effect that "perhaps it would not yet be safe to say that *all* forms," etc. Nay, not only does he directly *say* that "it is by no means his intention to suggest that there is no difference between the lowest plant and the highest, or between plants and animals," but he directly proves what he says, for he demonstrates in plants and animals an *essential difference of power.* Plants *can* assimilate inorganic matters, animals can *not*, etc. Again, here is a passage in which he is seen to cut his own "*basis*" from beneath his own feet. After telling us that all forms of protoplasm consist of carbon, hydrogen, oxygen, and nitrogen "in very complex union," he continues, "To this complex combination, *the nature of which has never been determined with exactness*, the name of protein has been applied." This, plainly, is an identification, on Mr. Huxley's own part, of protoplasm and protein ; and what is said of the one being necessarily true of the other, it follows that Mr. Huxley admits the nature of protoplasm never to have been determined with exactness, and that, even in his eyes, the *lis* is still *sub judice.* This admission is strengthened by the words, too, "If we use this term" (protein) "with such *caution* as may properly arise out of our *comparative ignorance* of the things for which it stands ;" which entitle us to recommend, in consequence "of our comparative ignorance of the things for which it stands," "*caution*" in the use of the term protoplasm. In such a state of the case we cannot wonder that Mr. Huxley's own conclusion here is : Therefore "all living matter is more or less albuminoid." All living matter is more or less albuminoid ! That, indeed, is the single conclusion of Mr. Huxley's whole industry ; but it is a conclusion that, far from requiring the intervention of protoplasm, had been reached long before the word itself had been, in this connection, used.

It is in this way, then, that Mr. Huxley can be ad-duced in refutation of himself ; and I think his resort to an epigram of Goethe's for reduction of the powers of man to those of contraction, digestion, and repro-duction, can be regarded as an admission to the same effect. The epigram runs thus :—

" Warum treibt sich das Volk so, und schreit ? Es will sich ernähren,
Kinder zeugen, und die nähren so gut es vermag.
Weiter bringt es kein Mensch, stell' er sich wie er auch will."

That means, quite literally translated, " Why do the folks bustle and bawl ? They want to feed themselves, get children, and then feed them as best they can ; no man does more, let him do as he may." This, really, is Mr. Huxley's sole proof for his classification of the powers of man. Is it sufficient ? Does it not apply rather to the birds of the air, the fish of the sea, and the beasts of the field, than to man ? Did Newton only feed himself, beget children, and then feed them ? Was it impossible for him to do any more, let him do as he might ? And what we ask of Newton we may ask of all the rest. To elevate, therefore, the passing whim of mere literary *Laune* into a cosmical axiom and a proof in place—this we cannot help adding to the other productions here in which Mr. Huxley appears against himself.

But were it impossible either for him or us to point to these *lacunæ*, it would still be our right and our duty to refer to the present conditions of microscopic science in general as well as in particular, and to demur to the erection of its *dicta*, constituted as they yet are, into established columns and buttresses in support of any theory of life, material or other.

The most delicate and dubious of all the sciences, it is also the youngest. In its manipulations the slightest change may operate as a destructive drought, or an equally destructive deluge. Its very tools may positively create the structure it actually examines. The present state of the science, and what warrant it gives Mr. Huxley to dogmatize on protoplasm, we may understand from this avowal of Kühne's : " To-day we believe that we see" such or such fact, " but know not that further improvements in the means of observation will not reveal what is assumed for certainty to be only illusion." With such authority to lean on—and it is the highest we can have—we may be allowed to entertain the conjecture, that it is just possible that some certainties, even of Mr. Huxley, may yet reveal themselves as illusions.

But, in resistance to any sweeping conclusions built on it, we are not confined to a reference to the imperfections involved in the very nature and epoch of the science itself in general. With yet greater assurance of carrying conviction with us, we may point in particular to the actual opinions of its present professors. We have seen already, in the consideration premised, that Mr. Huxley's hypothesis of a protoplasm *matter* is unsupported, even by the most innovating Germans, who as yet will not advance, the most advanced of them, beyond a protoplasm-cell ; and that his whole argument is thus sapped in advance. But what threatens more absolute extinction of this argument still, *all* the German physiologists do *not* accept even the protoplasm-cell. Rindfleisch, for example, in his recently-published 'Lehrbuch der pathologischen Gewebelehre,' speaks of the cell very much as we understand Virchow to have spoken of it. To him there is in the cell not only protoplasm but nucleus, and perhaps membrane as well. To him, too, the cell propagates itself quite as we have been hitherto fancying it to do, by division of the nucleus, increase of the protoplasm, and ultimate partition of the cell itself. Yet he knows withal of the opinions of others, and accepts them in a manner. He mentions Kühne's account of the membrane as at first but a mere physical limit of two fluids—a mere peripheral film or curdling ; still he assumes a formal and decided membrane at last. Even Leydig and Schultze, who shall be the express eliminators of the membrane —the one by initiation and the other by consummation— confess that, as regards the cells of certain tissues, they have never been able to detect in them the absence of a membrane.

As regards the nucleus again, the case is very much stronger. When we have admitted with Brücke that certain cryptogam cells, with Haeckel that certain protists, with Cienkowsky that two monads, and with Schultze that one amoeba, are without nucleus—when we have admitted that division of the cell *may* take place without implicating that of the nucleus—that the movements of the nucleus *may* be passive and due to those of the protoplasm—that Baer and Stricker demonstrate the disappearance of the original nucleus in the impregnated egg,—when we have admitted this, we have admitted also all that can be said in degradation

of the nucleus. Even those who say all this still attribute to the nucleus an important and unknown *rôle*,
and describe the formation in the impregnated egg of a
new nucleus ; while there are others again who resist
every attempt to degrade it. Böttcher asserts movement for the nucleus, even when wholly removed from
the cell ; Neumann points to such movement in dead
or dying cells ; and there is other testimony to a like
effect, as well as to peculiarities of the nucleus otherwise that indicate spontaneity. In this reference we
may allude to the weighty opinion of the late Professor
Goodsir, who anticipated in so remarkable a manner
certain of the determinations of Virchow. Goodsir, in
·that anticipation, wonderfully rich and ingenious as he
is everywhere, is perhaps nowhere more interesting and
successful than in what concerns the nucleus. Of the
whole cell, the nucleus is to him, as it was to Schleiden,
Schwann, and others, the most important element.
And this is the view to which I, who have little business to speak, wish success. This universe is not an
accidental cavity, in which an accidental dust has been
accidentally swept into heaps for the accidental evolution of the majestic spectacle of organic and inorganic
life. That majestic spectacle is a spectacle as plainly
for the eye of reason as any diagram of the mathematician. That majestic spectacle could have been constructed, was constructed, only in reason, for reason,
and by reason. From beyond Orion and the Pleiades,
across the green hem of earth, up to the imperial per-
sonality of man, all, the furthest, the deadest, the dus-
tiest, is for fusion in the invisible point of the single
Ego—*which alone glorifies it*. *For* the subject, and on
the model of the subject, all is made. Therefore it is

that—though, precisely as there are acephalous monsters by way of exception and deformity, there may be also at the very extremity of animated existence cells without a nucleus—I cannot help believing that this nucleus itself, as analogue of the subject will yet be proved the most important and indispensable of all the normal cell-elements. Even the phenomena of the impregnated egg seem to me to support this view. In the egg, on impregnation, it seems to me natural (I say it with a smile) that the old sun that ruled it should go down, and that a new sun, stronger in the combination of the new and the old, should ascend into its place !

Be these things as they may, we have now overwhelming evidence before us for concluding, with reference to Mr. Huxley's first proposition, that—in view of the nature of microscopic science—in view of the state of belief that obtains at present as regards nucleus, membrane, and entire cell—even in view of the supporters of protoplasm itself—Mr. Huxley is not authorized to speak of a physical matter of life ; which, for the rest, if granted, would, for innumerable and, as it appears to me, irrefragable reasons, be obliged to acknowledge for itself, not identity, but an infinite diversity .in power, in form and in substance.

So much for the first proposition in Mr. Huxley's essay, or that which concerns protoplasm, as a supposed matter of life, identical itself, and involving the identity of all the various organs and organisms which it is assumed to compose. What now of the second proposition, or that which concerns the materiality at once of protoplasm, and of all that is conceived to derive from protoplasm ? In other words, though, so to speak, for organic bricks anything like an organic clay still awaits the proof, I ask, if the bricks are not the same because the clay is not the same, what if the materiality of the former is equally unsupported by the materiality of the latter ? Or what if the functions of protoplasm are not properties of its mere molecular constitution ?

For this is Mr. Huxley,s second proposition, namely, That. all vital and intellectual functions are but the properties of the molecular disposition and changes of the material basis (protoplasm) of which the various animals and vegetables consist. With the conclusions now before us, it is evident that to enter at all on this part of Mr. Huxley's argumentation is, so far as we are concerned, only a matter of grace. In order that it should have any weight, we must grant the fact, at once of the existence of a matter of life, and of all organs and organisms being but aggregates of it. This, obviously, we cannot now do. By way of hypothesis, however, we may assume it. Let it be granted, then, that *pro hac vice* there *is* a physical basis of life with all the consequences named ; and now let us see how Mr. Huxley proceeds to establish its materiality.

The whole former part of Mr. Huxley's essay consists (as said) of fifty paragraphs, and the argument immediately concerned is confined to the latter ten of them. This argument is the simple chemical analogy that, under stimulus of an electric spark, hydrogen and oxygen uniting into an equivalent weight of water, and, under stimulus of preëxisting protoplasm, carbon, hydrogen, oxygen, and nitrogen uniting into an equivalent weight of protoplasm, there is the same warrant for atttributing the properties of the consequent to the properties of the antecedents in the latter case as in the former. The properties of protoplasm are, in origin and charac-

ter, precisely on the same level as the properties of water. The cases are perfectly parallel. It is as absurd
to attribute a new entity vitality to protoplasm, as a new
entity aquosity to water. Or, if it is by its mere chemical and physical structure that water exhibits certain
.properties called aqueous, it is also by its mere chemical and physical structure that protoplasm exhibits certain properties called vital. All that is necessary in
either case is, "under certain conditions," to bring the
chemical constituents together. If water is a molecular complication, protoplasm is equally a molecular complication, and for the description of the one or the
other there is no change of language required. A new
substance with new qualities results in precisely the
same way here, as a new substance with new qualities ·
there ; and the derivative qualities are not more different from the primitive qualities in the one instance,
than the derivative qualities are different from the primitive qualities in the other. Lastly, the *modus operandi*
of preëxistent protoplasm is not more unintelligible than,
that of the electric spark. The conclusion is irresistible, then, that all protoplasm being reciprocally convertible, and consequently identical, the properties it
displays, vitality and intellect included, are as much
the result of molecular constitution as those of water
itself.

It is evident, then, that the fulcrum on which Mr.
Huxley's second proposition rests, is a single inference
from a chemical analogy. Analogy, however, being
never identity, is apt to betray. The difference it hides
may be essential, that is, while the likeness it shows
may be inessential—so· far as the conclusion is concerned. That this mischance has overtaken Mr. Hux-

ley here, it will, I fancy, not be difficult to demonstrate.

The analogy to which Mr. Huxley trusts has two references: one, to chemical composition, and one to a certain stimulus that determines it. As regards chemical composition, we are asked, by virtue of the analogy obtaining, to identify, as equally simple instances of it, protoplasm here and water there ; and, as regards the stimulus in question, we are asked to admit the action of the electric spark in the one case to be quite analogous to the action of preëxisting protoplasm in the other. In both references I shall endeavor to point out that the analogy fails ; or, as we may say it also, that, even to Mr. Huxley, it can only seem to succeed by discounting the elements of difference that still subsist.

To begin with chemical combination, it is not unjust to demand that the analogy which must be admitted to exist in that, and a general physical respect, should not be strained beyond its legitimate limits. Protoplasm cannot be denied to be a chemical substance ; protoplasm cannot be denied to be a physical substance. As a compound of carbon, hydrogen, oxygen and nitrogen, it comports itself chemically—at least in ultimate instance—in a manner not essentially different from that in which water, as a compound of hydrogen and oxygen, comports itself chemically. In mere physical aspect, again, it may count quality for quality with water in the same aspect. In short, so far as it is on chemical and physical structure that the possession of distinctive properties in any case depends, both bodies may be allowed to be pretty well on a par. The analogy must be allowed to hold so far: so far but no farther. One step farther and we see not only that protoplasm has, like water, a chemical and physical structure ; but that, unlike water, it has also an organized or organic structure. Now this, on the part of protoplasm, is a possession in excess ; and with relation to that excess there can be no grounds for analogy. This, perhaps, is what Mr. Huxley has omitted to consider. When insisting on attributing to protoplasm the qualities it possessed, because of its chemical and physical structure, if it was for chemical and physical structure that we attributed to water *its* qualities, he has simply forgotten the addition to protoplasm of a third structure that can only be named organic. " If the phenomena exhibited by water are its properties, so are those presented by protoplasm, living or dead, its properties." When Mr. Huxley speaks thus, Exactly so, we may answer : " living or dead !" That alternative is simply slipped in and passed ; but it is in that alternative that the whole matter lies. Chemically, dead protoplasm is to Mr. Huxley quite as good as living protoplasm. As a sample of the article, he is quite content with dead protoplasm, and even swallows it, he says, in the shape of bread, lobster, mutton, etc., with all the satisfactory results to be desired. - Still, as concerns the argument, it must be pointed out that it is only these that can be placed on the same level as water ; and that living protoplasm is not only unlike water, but it is unlike dead protoplasm. Living protoplasm, namely, is identical with dead protoplasm only so far as its chemistry is concerned (if even so much as that) ; and it is quite evident, consequently, that difference between the two cannot depend on that in which they are identical—cannot depend on the chemistry. Life, then, is no affair of chemical and physical structure, and must find its explanation in something else. It is thus that, lifted high enough, the light of the analogy between water and protoplasm is seen to go out. Water, in fact, when formed from hydrogen and oxygen, is, in a certain way and in relation to them, no new product ; it has still, like them, only chemical and physical qualities ; it is still, as they are, inorganic. So far as *kind* of power is concerned, they are still on the same level. But not so protoplasm, where, with preservation of the chemical and physical likeness there is the addition of the unlikeness of life, of organization, and of ideas. But the addition is a new world—a new and higher world, the world of a self-realizing thought, the world of an *entelechy*. The change of language objected to by Mr. Huxley is thus a matter of necessity, for it is *not* mere molecular complication that we have any longer before us, and the qualities of the derivative are essentially and absolutely different from the qualities of the primitive. If we did invent the term aquosity, then, as an abstract sign for all the qualities of water, we should really do very little harm ; but aquosity and vitality would still remain essentially unlike. While for the invention of aquosity there is little or no call, however, the fact in the other case is that we are not only compelled to invent, but to *perceive* vitality. We are quite willing to do as Mr. Huxley would have us to do : look on, watch the phenomena, and name the results. But just in proportion to our faithfulness in these respects is the necessity for the recognition of a new world and a new nomenclature. There are certainly different states of water, as ice and steam ; but the relation of the solid to the liquid, or of either to the vapor, surely offers no analogy to the relation of protoplasm dead to protoplasm alive. That relation is not an analogy but an antithesis, the antithesis of antitheses. In it, in fact, we are in presence of the one incommunicable gulf—the gulf of all gulfs—that gulf which Mr. Huxléy's protoplasm is as powerless to efface as any other material expedient that has ever been suggested since the eyes of men first looked into it—the mighty gulf between death and life.

The differences alluded to (they are, in order, organization and life, the objective idea—design, and the subjective idea—thought), it may be remarked, are admitted by those very Germans to whom protoplasm, name and thing, is due. They, the most advanced and innovating of them, directly avow that there is present in the cell " an architectonic principle that has not yet been detected." In pronouncing protoplasm capable of active or vital movements, they do by that refer, they admit also, to an immaterial force, and they ascribe the processes exhibited by protoplasm—in so many words— not to the molecules, but to organization and life. It is remarked by Kant that " the reason of the specific mode of existence of every part of a living body lies in the whole, whilst with dead masses each part bears this reason within itself;" and this indeed is how the two worlds are differentiated. A drop of water, once formed, is there passive for ever, susceptible to influence, but indifferent to influence, and what influence reaches it is wholly from without. It may be added to, it may be subtracted from ; but infinitely apathetic quantitatively, it is qualitatively independent. It is indifferent to its own physical parts. It is without contractility, without alimentation, without reproduction, without specific function. Not so the cell, in which the parts are dependent on the whole, and the whole on the parts; which has its activity and *raison d'être* within ; which manifests all the powers which we have described water to want ; and which requires for its continuance conditions of which water is independent. It is only so far as organization and life are concerned, however, that the cell is thus different from water. Chemically and physically, as said, it can show with it quality for quality. How strangely Mr. Huxley's deliverances show beside these facts ! He can " see no break in the series of steps in molecular complication ;" but, glaringly obvious, there is a step added that is not molecular at all, and that has its supporting conditions completely elsewhere. The molecules are as fully accounted for in protoplasm as in water ; but the sum of qualities, thus exhausted in the latter, is not so exhausted in the former, in which there are qualities due, plainly, not to the molecules as molecules, but to the form into which they are thrown, and the force that makes that form one. When the chemical elements are brought together, Mr. Huxley says, protoplasm is formed, " and this protoplasm exhibits the phenomena of life ;" but he ought to have added that these phenomena are themselves added to the phenomena for which all that relates to chemistry stands, and are there, consequently, only by reason of some other determinant. New consequents necessarily demand new antecedents. " We think fit to call different kinds of matter carbon, oxygen, hydrogen, and nitrogen, and to speak of the various powers and activities of these substances as the properties of the matter of which they are composed." That, doubtless, is true, we say ; but such statements do not exhaust the facts. We call water hydrogen and oxygen, and attribute *its* properties· to the properties of them. In a chemical point of view, we ought to do the same thing for ice and steam ; yet, for all the chemical identity, water is not ice, nor is either steam. Do we, then, in these cases, make nothing of the *difference*, and in its despite enjoy the satisfaction of viewing the three as one ? Not so ; we ask a reason for the difference ; we demand an antecedent that shall render the consequent intelligible. The chemistry of oxygen and hydrogen is not enough in explanation of the threefold form ; and by the very necessity of the facts we, are driven to the addition of heat. It is precisely so with protoplasm in its twofold form. The chemistry remaining the same in each (if it really does so), we are compelled to seek elsewhere a reason for the difference of living from dead protoplasm. As the differences of ice and steam from water lay not in the hydrogen and oxygen, but in the heat, so the difference of living from dead protoplasm lies not in the carbon, the hydrogen, the oxygen, and the nitrogen, but in the vital organization. In all cases, for the new quality, plainly, we must have a new explanation. The qualities of a steam-engine are not the results of its simple chemistry. We do apply to protoplasm the same conceptions, then, that are legitimate elsewhere, and in allocating properties and explaining phenomena we simply insist on Mr. Huxley's own distinction of " living or dead." That, in fact, is to us the distinction of distinctions, and we admit no vital action whatever, not even the dullest, to be the result of the *molecular* action of the protoplasm that displays it. The very protoplasm of the nettie-sting, with which Mr. Huxley begins, is already vitally organized, and in that organization as much superior to its own molecules as the steam-engine, in its mechanism, to its own wood and iron. It were indeed as rational to say that there is no principle concerned in a steam-engine or a watch but that of its molecular forces, as to make this assertion of organized matter. Still there are degrees in organization, and the highest forms of life are widely different from the lowest. Degrees similar we see even in the inorganic world. The persistent flow of a river is, to the mighty reason of the solar system, in some such proportion, perhaps, as the rhizopod to man. In protoplasm, even the lowest, then, but much more conspicuously in the highest, there is, in addition to the molecular force, another force unsignalized by Mr. Huxley—the force of vital organization.

But this force is a rational unity, and that is an idea ; and this I would point to as a second form of the addition to the chemistry and physics of protoplasm. We have just seen, it is true, that an idea may be found in inorganic matter, as in the solar and sidereal systems generally. But the idea in organized matter is not one operative, so to speak, from without : it is one operative from within, and in an infinitely more intimate and pervading manner. The units that form the complement of an inorganic system are but independently and externally in place, like units in a procession ; but in what is organized there is no individual that is not sublated into the unity of the single life. This is so even in protoplasm. Mr. Huxley, it is true, desiderates, as result of mere ordinary chemical process, a life-stuff in mass, as it were in the web, to which he has only to resort for cuttings and cuttings in order to produce, by aggregation, what organized individual he pleases. But the facts are not so : we cannot have protoplasm in the web, but the piece. There is as yet no *matter* of life ; there are still *cells* of life. It is no shred of protoplasm —no spoonful or toothpickful—that can be recognized as adequate to the function and the name. Such shred may wriggle a moment, but it produces nought, and it dies. In the smallest, lowest protoplasm cell, then, we have this rational unity of a complement of individuals that only are for the whole and exist in the whole. This is an idea, therefore; this is design : the organized concert of many to a single common purpose. The rudest savage that should, as in Paley's illustration, find a watch, and should observe the various contrivances all controlled by the single end in view, would be obliged to acknowledge—though in his own way—that what he had before him was no mere physical, no mere molecular product. So in protoplasm : even from the first, but, quite undeniably, in the completed organization at last, which alone it was there to produce ; for a single idea has been its one manifestation throughout. And in what machinery does it not at length issue ? Was it molecular powers that invented a respiration— that perforated the posterior ear to give a balance of air—that compensated the *fenestra ovalis* by a *fenestra rotunda*—that placed in the auricular sacs those *otolithes*, those express stones for hearing ? Such machinery ! The *chordæ tendineæ* are to the valves of the heart exactly adjusted check-strings ; and the contractile *columnæ carneæ* are set in, under contraction and expansion, to equalize their length to their office. Membranes, rods, and liquids—it required the express experiment of man to make good the fact that the inventor of the ear had availed himself of the most perfect apparatus possible for his purpose. And are we

to conceive such machinery, such apparatus, such contrivances merely molecular? Are molecules adequate
to such things—molecules in their blind passivity, and
dead, dull insensibility? Is it to molecular agency Mr.
Huxley himself owes that " singular inward laboratory"
of which he speaks, and without which all the protoplasm in the world would be useless to him? Surely,
in the presence of these manifest ideas, it is impossible
to attribute the single peculiar feature of protoplasm—
its vitality, namely—to mere molecular chemistry. Protoplasm, it is true, breaks up into carbon, hydrogen,
oxygen, and nitrogen, as water does into hydrogen and
oxygen ; but the watch breaks similarly up into mere
brass, and steel, and glass. The loose materials of
the watch—even its chemical material if you will—replace its weight, quite as accurately as the constituents
carbon, etc., replace the weight of the protoplasm.
But neither these nor those replace the vanished idea,
which was alone the important element. Mr. Huxley
saw no break in the series of steps in molecular complication ; but, though not molecular, it is difficult to
understand what more striding, what more absolute
break could be desired than the break into an idea. It
is of that break alone that we think in the watch ; and
it is of that break alone that we should think in the
protoplasm which, far more cunningly, far more ration-
ally, constructs a heart, an eye or an ear. That is the
break of breaks, and explain it as we may, we shall
never explain it by molecules.

But, if inorganic elements as such are inadequate to
account either for vital organization or the objective
idea of design, much more are they inadequate, in the
third place, to account for the subjective idea, for the

phenomena of thought as thought. Yet Mr. Huxley tells us that thought is but the expression of the molecular changes of protoplasm. This he only tells us ; this he does not prove. He merely says that, if we admit the functions of the lowest forms of life to be but "direct results of the nature of the matter of which they are composed," we must admit as much for the functions of the highest. We have not admitted Mr. Huxley's presupposition ; but, even with its admission, we should not feel bound to admit his conclusion. In such a mighty system of differences, there are ample room and verge enough for the introduction of new motives. We can say here at once, in fact, that as thought, let its connection be what it may with, has never been proved to result from, organization, no improvement of the proof required will be found in protoplasm. No one power that Mr. Huxley signalizes in protoplasm can account for thought : not alimentation, and not reproduction, certainly ; but not even contractility. We have seen already that there is no proof of contraction being necessary even for the simplest sensation ; but much less is there any proof of a necessity of contraction for the inner and independent operations of the mind. Mr. Huxley himself admits this. He says : " Speech, gesture, and every other form of human action are, in the long-run, resolvable into muscular contraction ;" and so, " even those manifestations of intellect, of feeling, and of will, which we rightly name the higher faculties, are not excluded from this classification, inasmuch as to every one *but the subject of them*, they are known only as transitory changes in the relative positions of parts of the body." The concession is made here, we see, that these manifestations are differently known to the sub-

3

ject of them.   But we may first object that, if even that privileged " every one but the subject" were limited to a knowledge of contractions, he would not know much. It is only because he knows, first of all, a thinker and willer of contractions that these themselves cease to be but passing externalities, and transitory contingencies. Neither is it reasonable to assert an identity of nature for contractions, and for that which they only represent. It would hardly be fair to confound either the receiver or the sender of a telegraphic message, with the movements which alone bore it, and without which it would have been impossible.   The sign is not the thing signified, it is but the servant of the signifier—his own arbitrary mark—and intelligible, in the first place, only to him.   It is the meaning, in all cases, that is alone vital ; the sign is but an accident.   To convert the internality into the arbitrary externality that simply expresses it, is for Mr. Huxley only an oversight.   Your ideas are made known to your neighbors by contractions, therefore your ideas are of the same nature as contractions ! Or, even to take it from the other side, your neighbor perceives in you contractions only, and therefore your ideas are contractions !   Are not the vital elements here present the two correspondent internalities, between which the contractions constitute but an arbitrary chain of external communication, that is so now, but may be otherwise again ?   The ringing of the bell at the window is not precisely the dwarf within.   Nor are Engineer Chappe's " wooden arms and elbow-joints jerking and fugling in the air," to be identified with Engineer Chappe himself.   For the higher faculties, even for speech, etc., assuredly Mr. Huxley might have well spared himself this superfluous and inapplicable reference to contraction.

But, in the middle of it, as we have seen, Mr. Huxley concedes that these manifestations are differently known to the subject of them. If so, what becomes of his assertion of but a certain number of powers for protoplasm? The manifestations of the higher faculties are not known to the subject of them by contraction, etc. By what, then, are they known? According to Mr. Huxley, they can only be known by the powers of protoplasm; and therefore, by his own showing, protoplasm must possess powers other than those of his own assertion. Mr. Huxley's one great power of contractility, Mr. Huxley himself confesses to be inapplicable here. Indeed, in his Physiology (p. 193), he makes such an avowal as this: "We class *sensations*, along with *emotions*, and *volitions*, and *thoughts*, under the common head of states of *consciousness;* but what consciousness is we know not, and how it is that anything so remarkable as a state of consciousness comes about as the result of irritating nervous tissue, is just as unaccountable as the appearance of the Djin when Aladdin rubbed his lamp in the story." Consciousness plainly was not muscular contraction to Mr. Huxley when he wrote his Physiology; it is only since then that he has gone over to the assertion of no power in protoplasm but the triple power, contractility, etc. But the truth is only as his Physiology has it—the cleft is simply, as Mr. Huxley acknowledges it there, absolute. On one side, there is the world of externality, where all is body by body, and away from one another—the boundless reciprocal exclusion of the infinite object. On the other side, there is the world of internality, where all is soul to soul, and away into one another—the boundless reciprocal inclusion of the infinite subject. This—even

while it is true that, for subject to be subject, and object, object, the boundless intussuscepted multiplicity
of the single invisible point of the one is but the dimen-
sionless casket into which the illimitable Genius of the
other must retract and withdraw itself—is the difference of differences ; and certainly it is not internality
that can be abolished before externality. The proof
for the absoluteness of thought, the subject, the mind,
is, on its side, pretty well perfect. It is not necessary
here, however, to enter into that proof at length. Before passing on, I may simply point to the fact that, if
thought is to be called a function of matter, it must be
acknowledged to be a function wholly peculiar and unlike any other. In all other functions, we are present
to processes which are in the same sense physical as
the organs themselves. So it is with lung, stomach,
liver, kidney, where every step can be followed, so to
speak, with eye and hand ; but all is changed when we
have to do with mind as the function of brain. Then,
indeed, as Mr. Huxley thought in his Physiology, we are
admitted, as if by touch of Aladdin's lamp, to a world
absolutely different and essentially new—to a world, on
its side of the incommunicable cleft, as complete, entire, independent, self-contained, and absolutely *sui
generis*, as the world of matter on the other side. It
will be sufficient here to allude to as much as this, with
special reference to the fact that, so far as this argument is concerned, protoplasm has not introduced any
the very slightest difference. All the ancient reasons
for the independence of thought as against organization, can be used with even more striking effect as
against protoplasm ; but it will be sufficient to indicate
this, so much are the arguments in question a common

property now. Thought, in fact, brings with it its own warrant ; or it brings with it, to use the phrase of Burns, " its patent of nobility direct from Almighty God." And that is the strongest argument on this whole side. Throughout the entire universe, organic and inorganic, thought is the controlling sovereign ; nor does matter anywhere refuse its allegiance. So it is in thought, too, that man has *his* patent of nobility, believes that he is created in the image of God, and knows himself a free-man of infinitude.

But the analogy, in the hands of Mr. Huxley, has, we have seen, a second reference—that, namely, to the ex-citants, if we may call them so, which *determine* combi-nation. The *modus operandi*, Mr. Huxley tells us, of preëxisting protoplasm in determining the formation of new protoplasm, is not more unintelligible than the *modus operandi* of the electric spark in determining the formation of water ; and so both, we are left to infer, are perfectly analogous. The inferential turn here is rather a favorite with Mr. Huxley. " But objectors of this class," he says on an earlier occasion, in allusion to those who hesitate to conclude from dead to living matter, " do not seem to reflect that it is also, in strict-ness, true that we know nothing about the composition of any body whatever as it is." In the same neighbor-hood, too, he argues that, though impotent to restore to decomposed calc-spar its original form, we do not hesitate to accept the chemical analysis assigned to it, and should not, consequently, any more hesitate be-cause of any mere difference of form to accept the anal-ysis of dead for that of living protoplasm. It is cer-tainly fair to point out that, if we bear ignorance and impotence with equanimity in one case, we may equally so bear them in another ; but it is not fair to convert ignorance into knowledge, nor impotence into power. Yet it is usual to take such statements loosely, and let them pass. It is not considered that, if we know nothing about the composition of any body whatever as it is, then we do know nothing, and that it is strangely idle to offer absolute ignorance as a support for the most dogmatic knowledge. If such statements are, as is really expected for them, to be accepted, yet not accepted, they are the stultification of all logic. Is the chemistry of living to be seen to be the same as the chemistry of dead protoplasm, because we know nothing about the composition of any body whatever as it is ? We know perfectly well that black is white, for we are absolutely ignorant of either as it is ! The *form* of the calc-spar, which (the spar) we *can* analyze, we cannot restore ; therefore the *form* of the protoplasm, which we *cannot* analyze, has nothing to do with the matter in hand ; and the chemistry of what is dead may be accepted as the chemistry of what is living ! In the case of reasoning so irrelevant it is hardly worth while referring to what concerns the forms themselves ; that they are totally incommensurable, that in all forms of calc-spar there is no question but of what is physical, while in protoplasm the change of form is introduction into an entire new world. As in these illustrations, so in the case immediately before us. No appeal to ignorance in regard to something else, the electric spark, should be allowed to transform another ignorance, that of the action of preëxisting protoplasm, into knowledge, here into *the* knowledge that the two unknown things, because of non-knowledge, are—perfectly analogous ! That this analogy does not exist—that the electric spark and preëxisting protoplasm are, in their relative places, *not* on the same chemical level—this is the main point for us to see ; and Mr. Huxley's allusion to our ignorance must not be allowed to blind us to it. Here we have in a glass vessel so much hydrogen and oxygen, into which we discharge an electric spark, and water is the result. Now what analogy is it possible to perceive between this production of water by external experiment and the production of protoplasm by protoplasm ? The discrepancy is so palpable that it were impertinent to enlarge on it. The truth is just this, that the measured and mixed gases, the vessel, and the spark, in the one case, are as unlike the fortuitous food, the living organs, and the long process of assimilation in the other case, as the product water is unlike the product protoplasm. No ; that the action of the electric spark should be unknown, is no reason why we should not insist on protoplasm for protoplasm, on life for life. Protoplasm can only be produced by protoplasm, and each of all the innumerable varieties of protoplasm, only by its own kind. For the protoplasm of the worm we must go to the worm, and for that of the toad-stool to the toad-stool. In fact, if all living beings come from protoplasm, it is quite as certain that, but for living beings, protoplasm would disappear. Without an egg you cannot have a hen—that is true ; but it is equally true that, without a hen, you cannot have an egg. So in protoplasm ; which, consequently, in the production of itself, offers no analogy to the production, or precipitation by the electric spark, not of itself, but of water. Besides, if for protoplasm, preëxisting protoplasm, is always necessary, how was there ever a first protoplasm ?

Generally, then, Mr. Huxley's analogy does not hold, whether in the one reference or the other, and Mr. Huxley has no warrant for the reduction of protoplasm to the mere chemical level which he assigns it in either. That level is brought very prominently forward in such expressions as these : That it is only necessary to bring the chemical elements " together," " under certain conditions," to give rise to the more complex body, protoplasm, just as there is a similar expedient to give rise to water ; and that, under the influence of pre-existing living protoplasm, carbonic acid, water, and ammonia disappear, and an equivalent weight of protoplasm makes its appearance, just as, under the influence of the electric spark, hydrogen and oxygen disappear, and an equivalent weight of water makes its appearance. All this, plainly, is to assume for protoplasm such mere chemical place and nature as consist not with the facts. The cases are, in truth, not parallel, and the " certain conditions" are wholly diverse. All that is said we can do at will for water, but nothing of what is said can we do at will for protoplasm. To say we can feed protoplasm, and so make protoplasm at will produce protoplasm, is very much, in the circumstances, only to say, and is not to say, that, in this way, we make a chemical experiment. To insist on a chemical analogy, in fact, between water and protoplasm, is to omit the differences not covered by the analogy at all— thought, design, life, and all the processes of organization ; and it is but simple procedure to omit these differences only by an appeal to ignorance elsewhere.

It is hardly worth while, perhaps, to refer now again to the difference—here, however, once more incidentally suggested—between protoplasm and protoplasm.

Mr. Huxley, that is, almost in his very last word on this part of the argument, seems to become aware of the bearing of this on what relates to materiality, and he would again stamp protoplasm (and with it life and intellect), into an indifferent identity. In order that there should be no break between the lowest functions and the highest (the functions of the fungus and the functions of man), he has " endeavored to prove," he says, that the protoplasm of the lowest organisms is "essentially identical with, and most readily converted into that of any animal." On this alleged reciprocal *convertibility* of protoplasm, then, Mr. Huxley would again found as well an inference of identity, as the further conclusion that the functions of the highest, not less than those of the lowest animals, are but the molecular manifestations of their common protoplasm.

Plainly here it is only the consideration, not of function, but of the alleged reciprocal *convertibility* that is left us now. Is this true, then? Is it true that every organism can digest every other organism, and that thus a relation of identity is established between that which digests and whatever is digested? These questions place Mr. Huxley's general enterprise, perhaps, in the most glaring light yet ; for it is very evident that there is an end of the argument if all foods and all feeders are essentially identical both with themselves and with each other. The facts of the case, however, I believe to be too well known to require a single word here on my part. It is not long since Mr. Huxley himself pointed out the great difference between the foods of plants and the foods of animals ; and the reader may be safely left to think for himself of *ruminantia* and *carnivora*, of soft bills and hard bills, of molluscs and men. Mr. Huxley talks feelingly of the possibility of himself feeding the lobster quite as much as of the lobster feeding him ; but such pathos is not always applicable ; it is not likely that a sponge would be to the stomach of Mr. Huxley any more than Mr. Huxley to the stomach of a sponge.

But a more important point is this, that the functions themselves remain quite apart from the alleged convertibility. We can neither acquire the functions of what we eat, nor impart our functions to what eats us. We shall not come to fly by feeding on vultures, nor they to speak by feeding on us. No possible manure of human brains will enable a corn-field to reason. But if functions are inconvertible, the convertibility of the protoplasm is idle. In this inconvertibility, indeed, functions will be seen to be independent of mere chemical composition. And that is the truth : for functions there is more required than either chemistry or physics.

It is to be acknowledged—to notice one other incidental suggestion, for the sake of completeness, and by way of transition to the final consideration of possible objections—that Mr. Huxley would be very much assisted in his identification of differences, were but the theories of the molecularists, on the one hand, and of Mr. Darwin, on the other, once for all established. The three modes of theorizing indicated, indeed, are not without a tendency to approach one another ; and it is precisely their union that would secure a definitive triumph for the doctrine of materialism. Mr. Huxley, as we have seen—though what he desiderates is an autoplastic living *matter* that, produced by ordinary chemical processes, is yet capable of continuing and developing itself into new and higher forms—still begins with

the egg. Now the theory of the molecularists would,
for its part, remove all the difficulties that, for materialism, are involved in this beginning ; it would place protoplasm undeniably at length on a merely chemical
level ; and would fairly enable Mr. Darwin, supple-
mented by such a' life-stuff, to account by natural means
for everything like an idea or thought that appears in
creation. The misfortune is, however, that we must
believe the theory of the molecularists still to await the
proof ; while the theory of Mr. Darwin has many difficulties peculiar to itself. This theory, philosophically,
or in ultimate analysis, is an attempt to prove that design, or the objective idea, especially in the organic
world, is developed *in time* by natural means. The time
which Mr. Darwin demands, it is true, is an infinite
time ; and he thus gains the advantage of his processes
being allowed greater *clearness* for the understanding, in
consequence of the *obscurity* of the infinite past in
which they are placed, and of which it is difficult in the
first instance to deny any possibility whatever. Still it
remains to be asked, Are such processes credible in any
time ? What Mr. Darwin has done in aid of his view
is, first, to lay before us a knowledge of facts in natural
history of surprising richness ; and, second, to support
this knowledge by an inexhaustible ingenuity of hypothesis in arrangement of appearances. Now, in both respects, whether for information or even interest, the
value of Mr. Darwin's contribution will probably always
remain independent of the argument or arguments that
might destroy his leading proposition ; and it is with
this proposition that we have here alone to do. As
said, we ask only, Is it true that the objective idea, the
design which we see in the organized world, is the re-

sult in infinite time of the necessary adaption of living structures to the peculiarities of the conditions by which they are surrounded?

Against this theory, then, its own absolute generalization may be viewed as our first objection. In ultimate abstraction, that is, the only agency postulated by Mr. Darwin is time—infinite time; and as regards actually existent beings and actually existent conditions, it is hardly possible to deny any possibility whatever to infinitude. If told, for example, that the elephant, if only obliged *infinitely* to run, might be converted into the stag, how should we be able to deny? So also, if the lengthening of the giraffe's neck were hypothetically attributed to a succession of dearths in infinite time that only left the leaves of trees for long-necked animals to live on, we should be similarly situated as regards denial. Still it can be pointed out that ingenuity of natural conjecture has, in such cases, no less wide a field for the negation than for the affirmation; and that, on the question of fact, nothing is capable of being determined. But we can also say more than that— we can say that any fruitful application even of *infinite time* to the *general problem of difference* in the world is inconceivable. To explain all from an absolute beginning requires us to commence with nothing; but to this nothing time itself is an addition. Time is an entity, a something, a difference added to the original identity: whence or how came time? Time cannot account for its own self; how is it that there is such a thing as time? Then no conceivable brooding even of infinite time could hatch the infinitude of space. How is it there is such a thing as space? No possible clasps of time and space, further, could ever conceivably thicken into matter. How is it there is such a thing as matter ? Lastly, so far, no conceivable brooding, or even gyrating, of a single matter in time and space could account for the specification of matter—carbon, gold, iodine, etc.—as we see and know it. Time, space, matter, and the whole inorganic world, thus remain impassive to the action even of infinite time ; all *these* differences remain incapable of being accounted for so.

But suppose no curiosity had ever been felt in this reference, which, though scientifically indefensible, is quite possible, how about the transition of the inorganic into the organic ? Mr. Huxley tells us that, for food, the plant needs nothing but its bath of smelling-salts. Suppose this bath now—a pool of a solution of carbonate of ammonia ; can any action of sun, or air, or electricity, be conceived to develop a cell—or even so much lump-protoplasm—in this solution ? The production of an initial cell in any such manner will not allow itself to be realized to thought. Then we have just to think for a moment of the vast differences into which, for the production of the present organized world, this cell must be distributed, to shake our heads and say we cannot well refuse anything to an infinite time, but still we must pronounce a problem of this reach hopeless.

It is precisely in conditions, however, that Mr. Darwin claims a solution of this problem. Conditions concern all that relates to air, heat, light, land, water, and whatever they imply. Our second objection, consequently, is, that conditions are quite inadequate to account for present organized differences, from a single cell. Geological time, for example, falls short, after all, of infinite time ; or, in known geological eras, let us calculate them as liberally as we may, there is not time

enough to account for the presently-existing varieties,
from one, or even several, primordial forms. So to
speak, it is not *in* geological time to account for the
transformation of the elephant into the stag from acceleration, or for that of the stag into the elephant from
retardation, of movement. And we may speak similarly of the growth of the neck of the giraffe, or even
of the elevation of the monkey into man. Moreover,
time apart, conditions have no such power in themselves.
It is impossible to conceive of animal or vegetable
effluvia ever creating the nerve by which they are felt,
and so gradually the Schneiderian membrane, nose, and
whole olfactory apparatus. Yet these effluvia are the
conditions of smell, and, *ex hypothesi*, ought to have
created it. Did light, or did the pulsations of the air,
ever by any length of time, indent into the sensitive
cell, eyes, and a pair of eyes—ears, and a pair of ears?
Light conceivably might shine for ever without such a
wonderfully complicated result as an eye. Similarly,
for delicacy and marvellous ingenuity of structure, the
ear is scarcely inferior to the eye ; and surely it is possible to think of a whole infinitude of those fitful and
fortuitous air-tremblings, which we call sound, without
indentation into anything whatever of such an organ.

A third objection to Mr. Darwin's theory is, that the
play of natural contingency in regard to the vicissi-
tudes of conditions, has no title to be named *selection*.
Naturalists have long known and spoken of the "influence of accidental causes ; but Mr. Darwin was the
first to apply the term *selection* to the action of these,
and thus convert accident into design. The agency to
which Mr. Darwin attributes all the changes which he
would signalize in animals is really the fortuitous con-

tingency of brute nature ; and it is altogether fallacious
to call such process, or such non-process, by a term involving foresight and a purpose. We have here, indeed,
only a metaphor wholly misapplied. The German writer who, many years ago, said "even the *genera* are
wholly a prey to the changes of the external universal
life," saw precisely what Mr. Darwin sees, but it never
struck him to style contingency selection. Yet, how
dangerous, how infectious, has not this ungrounded
metaphor proved ! It has become a *principle*, a *law*, and
been transferred by very genuine men into their own
sciences of philology and what not. People will wonder at all this by-and-by. But to point out the inappli-
cability of such a word to the processes of nature referred to by Mr. Darwin, is to point out also the impossibility of any such contingencies proceeding, by
graduated rise, from stage to stage, into the great symmetrical organic system—the vast plan—the grand harmonious whole—by which we are surrounded. This
rise, this system, is really the objective idea ; but it is
utterly incapable of being accounted for by any such •
agency as natural contingency in geological, or infinite,
or any time. But it is this which the word selection
tends to conceal.

We may say, lastly, in objection, here, that, in the fact
of " reversion" or " atavism," Mr. Darwin acknowledges
his own failure. We thus see that the species as species is something independent, and holds its own *insita
vis naturæ* within itself.

Probably it is not his theory, then, that gives value
to Mr. Darwin's book ; nor even his ready ingenuity,
whatever interest it may lend : it is the material information it contains. The ingenuity, namely, verges

somewhat on that Humian expedient of natural conjecture so copiously exemplified, on occasion of a few
trite texts, in Mr. Buckle. But that natural conjecture
is always insecure, equivocal, and many-sided. It may
be said that ancient warfare, for example, giving victory
always to the personally ablest and bravest, must have
resulted in the improvement of the race ; or that, the
weakest being always necessarily left at home, the improvement was balanced by deterioration ; or that the
ablest were necessarily the most exposed to danger, and
so, etc., etc., according, to ingenuity *usque ad infinitum*.
Trustworthy conclusion is nôt possible to this method,
but only to the induction of facts, or to scientific demonstration.

Neither molecularists nor Darwinians, then, are able
to level out the difference between organic and inorganic,
or between genera and genera or species and species.
The differences persist despite of both ; the distributed
identity remains unaccounted for. Nor, consequently,
is Mr. Darwin's theory competent to explain the objective idea by any reference to time and conditions. Living beings do exist in a mighty chain from the moss to
the man ; but that chain, far from founding, is founded
in the idea, and is not the result of any mere natural
growth of this into that. That chain is itself the most
brilliant stamp, the sign-manual, of design. On every
ledge of nature, from the lowest to the highest, there is
a life that is *its*,—a creature to represent it, reflect it—
so to speak, pasture on it. The last, highest, brightest
link of this chain is man ; the incarnation of thought itself, which is the summation of this universe ; man, that
includes in himself all other links and their single secret
—the personified universe, the subject of the world.

Mr. Huxley makes but small reference to thought ; he only tucks it in, as it were, as a mere appendicle of course.

It may be objected, indeed—to reach the last stage in this discussion—that, if Mr. Huxley has not disproved the conception of thought and life "as a something which works through matter, but is independent of it," neither have we proved it. But it is easy for us to reply that, if "*independent of*" means here "*unconnected with*," we have had no such object. We have had no object whatever, in fact, but to resist, now the extravagant assertion that all organized. tissue, from the lichen to Leibnitz, is alike in faculty, and again the equally extravagant assertion that life and thought are but ordinary products of molecular chemistry. As regards the latter assertion, we have endeavored to show that the processes of vital organization (as self-production, etc.) belong to another sphere, higher than, and very different from, those of mechanical juxtaposition or chemical neutralization ; that life, then, is no mere product of matter as matter ; that if no life can be pointed to independent of matter, neither is there any life-stuff independent of life ; and that life, consequently, adds a new and higher force to chemistry, as chemistry a new and higher force to mechanics, etc. As for thought, the endeavor was to show that it was as independent on the one side as matter on the other, that it controlled, used, summed, and was the reason of matter. Thought, then, is not to be reached by any bridge from matter, that is a hybrid of both, and explains the connection. The relation of matter to mind is not to be explained as a transition, but as a *contrecoup*. In this relation, however, it is not the material, but the mental side, which the whole universe declares to be the dominant one.

· As regards any objection to the arguments which we have brought against the identity of protoplasm, again, these will lie in the phrase, probably, " difference not of kind, but degree," or in the word "modification." The " phrase" may be now passed, for generic or specific difference must be allowed in protoplasm, if not for the overwhelming reason that an infinitude of various kinds exist in it, each of which is self-productive and uninterchangeable with the rest, then for Mr. Huxley's own reason, that plants assimilate inorganic matter and animals only organic. As for the objection "modification," again, the same consideration of generic difference must prove fatal to it. This were otherwise, indeed, could but the molecularists and Mr. Darwin succeed in destroying generic difference ; but in this, as we have seen, they have failed. And this will be always so : who dogs identity, difference dogs him. It is quite a justifiable endeavor, for example, to point out the identity that obtains between veins and arteries on the one hand, as between these and capillaries on the other ; but all the time the difference is behind us ; and when we turn to look, we see, for circulation, the valves of the veins and the elastic coats of the arteries as opposed to one another, and, for irrigation, the permeable walls of the capillaries as opposed to both.

Generic differences exist then, and we cannot allow the word " modification" to efface them in the interest of the identity claimed for protoplasm. Brain-protoplasm is not bone-protoplasm, nor the protoplasm of the fungus the protoplasm of man. Similarly, it is very questionable how far the word " modification" will warrant us in regarding with Mr. Huxley the " ducts, fibres, pollen, and ovules" of the nettle as identical with the protoplasm of its sting. Things that originate alike may surely eventuate in others which, chemically and vitally, far from being mere modifications, must be pronounced totally different. Such eventuation must be held competent to 'what can only be named generic or specific difference. The "child" is only "*father* of the man"—it is not the man ; who, moreover, in the course of an ordinary life, we are told, has totally changed himself, not once, but many times, retaining at the last not one single particle of matter with which he set out. Such eventuations, whether called modifications or not, certainly involve essential difference. And so situated are the "ducts, fibres, pollen, and ovules" of the nettle, which, whether compared with the protoplasm of the nettle-sting, or with that in which they originated, must be held to here assumed, by their own actions, indisputable differences, physical, chemical, and vital, or in form, substance, and faculty.

Much, in fact, depends on definition here ; and, in reference to modification, it may be regarded as arbitrary when identity shall be admitted to cease and difference to begin. There are the old Greek puzzles of the Bald Head and the Heap, for example. How many grains, or how many hairs, may we remove before a heap of wheat is no heap, or a head of hair bald ? These concern quantity alone ; but, in other cases, bone, muscle, brain, fungus, tree, man, there is not only a quantitative, but a qualitative difference ; and in regard to such differences, the word modification can be regarded as but a cloak, under which identity is to be shuffled into difference, but remain identity all the same. The brick is but modified clay, Mr. Huxley intimates, bake it and paint it as you may ; but is the difference introduced by the baking and painting to be ignored? Is what Mr. Huxley calls the "artifice" not to be taken into account, leave alone the "potter?" The strong firm rope is about as exact an example of modification proper— modification of the weak loose hemp—as can well be found; but are we to exclude from our consideration the whole element of difference due to the hand and brain of man? Not far from Burn's Monument, on the Calton Hill of Edinburgh, there lies a mass of stones which is potentially a church, the former Trinity College Church. Were this church again realized, would it be fair to call it a mere modification of the previous stones? Look now to the egg and the full-feathered fowl. Chaucer describes to us the cock, "hight chaunteclere," that was to his "faire Pertelotte" so dear:—

"His comb was redder than the fine corall,
Embattled, as it were a castle-wall;
His bill was black, and as the jet it shone;
Like azure were his legges and his tone (toes);
His nailes whiter than the lilie flour,
And like the burned gold was his color."

Would it be even as fair to call this fine fellow— comb, wattles, spurs, and all—a modified yolk, as to call the church but modified stones? If, in the latter case, an element of difference, altogether undeniable, seems to have intervened, is not such intervention at least quite as well marked in the former? It requires but a slight analysis to detect that all the stones in question are marked and numbered; but will any analysis point out within the shell the various parts that only need arrangement to become the fowl? Are the men that may take the stones, and, in a re-erected Trinity College Church, realize anew the idea of its architect,

in any respect more wonderful than the unknown dis-
posers of the materials of the fowl? That what rea-
lizes the idea should, in the one case, be from without,
and, in the other, from within, is no reason for seeing
more modification and less wonder in the latter than the
former. There is certainly no more reason for seeing
the fowl in the egg, and as identical with the egg, than
for seeing a re-built Trinity College Church as identical
with its unarranged materials. A part cannot be taken
for the whole, whether in space *or in time.* Mr. Huxley
misses this. He is so absorbed in the identity out of
which, that he will not see the difference into which,
progress is made. As the idea of the church has the
stones, so the idea of the fowl has the egg, for its commencement. But to this idea, and in both cases, the
terminal additions belong, quite as much as the initial
materials. If the idea, then, add sulphur, phosphorus,
iron, and what not, it must be credited with these not
less than with the carbon, hydrogen, etc., with which it
began. It is not fair to mutter modification, as if it
were a charm to destroy all the industry of time. The
protoplasm of the egg of the fowl is no more the fowl
than the stones the church; and to identify, by juggle
of a mere word, parts in time and wholes in time so different, is but self-deception. Nay, in protoplasm, as we
have so often seen, difference is as much present at first
as at last. Even in its germ, even in its initial identity,
to call it so, protoplasm is already different, for it issues
in differences infinite.

Omission of the consideration of difference, it is to be
acknowledged, is not now-a-days restricted to Mr. Huxley. In the wonder that is usually expressed, for example, at Oken's *identification* of the skull with so many

vertebræ, it is forgot that there is still implicated the wonder which we ought to feel at the unknown power that could, in the end, so *differentiate* them. If the cornea of the eye and the enamel of the teeth are alike but modified protoplasm, we must be pardoned for thinking more of the adjective than of the substantive. Our wonder is how, for one idea, protoplasm could become one thing here, and, for another idea, another so different thing there. We are more curious about the modification than the protoplasm. In the difference, rather than in the identity, it is, indeed, that the wonder lies. Here are several thousand pieces of protoplasm ; analysis can detect no difference in them. They are to us, let us say, as they are to Mr. Huxley, identical in power, in form, and in substance ; and yet on all these several thousand little bits of apparently indistinguishable matter an element of difference so pervading and so persistent has been impressed, that, of them all, not one is interchangeable with another ! Each seed feeds its own kind. The protoplasm of the gnat will no more grow into the fly than it will grow into an elephant. Protoplasm is protoplasm : yes, but man's protoplasm is man's protoplasm, and the mushroom's the mushroom's. In short, it is quite evident that the word modification, if it would conceal, is powerless to withdraw, the difference ; which difference, moreover, is one of kind and not of degree.

This consideration of possible objections, then, is the last we have to attend to ; and it only remains to draw the general conclusion. All animal and vegetable organisms are alike in power, in form, and in substance, only if the protoplasm of which they are composed is similarly alike ; and the functions of all animal and vegetable organisms are but properties of the molecular affections of their chemical constituents, only if the functions of the protoplasm, of which they are composed, are but properties of the molecular affections of *its* chemical constituents.   In disproof of the affirmative in both clauses, there has been no object but to demonstrate, on the one hand, the infinite non-identity of protoplasm, and, on the other, the dependence of its functions upon other factors than its molecular constituents.

In short, the whole position of Mr. Huxley, that all organisms consist alike of the same life-matter, which life-matter is, for its part, due only to chemistry, must be pronounced untenable—nor less untenable the materialism he would found on it.

ON THE HYPOTHESIS OF EVOLUTION:

PHYSICAL AND METAPHYSICAL.

UNIVERSITY

ON THE

HYPOTHESIS OF EVOLUTION:

PHYSICAL AND METAPHYSICAL.

"Man shall not live by bread alone, but by every word that pro-
ceedeth out of the mouth of God shall man live."

There is apparently considerable repugnance in the
minds of many excellent people to the acceptance, or
even consideration, of the hypothesis of development,
or that of the gradual creation by descent, with modification from the simplest beginnings, of the different
forms of the organic world. This objection probably
results from two considerations: first, that the human
species is certainly involved, and man's descent from
an ape asserted; and, secondly, that the scheme in
general seems to conflict with that presented by the
Mosaic account of the Creation, which is regarded as
communicated to its author by an infallible inspiration.

As the truth of the hypothesis is held to be infinitely
probable by a majority of the exponents of the natural
sciences at the present day, and is held as absolutely
demonstrated by another portion, it behooves those interested to restrain their condemnation, and on the
other hand to examine its evidences, and look any consequent necessary modification of our metaphysical or
theological views squarely in the face.

The following pages state a few of the former; if they suggest some of the latter, it is hoped that they may be such as any logical mind would deduce from the premises. That they will coincide with the spirit of the most advanced Christianity, I have no doubt; and that they will add an appeal through the reason to that direct influence of the Divine Spirit which should control the motives of human action, seems an unavoidable conclusion.

I. PHYSICAL EVOLUTION.

It is well known that a species is usually represented by a great number of individuals, distinguished from all other similar associations by more or less numerous points of structure, color, size, etc., and by habits and instincts also, to a certain extent; that the individuals of such associations reproduce their like, and cannot be produced by individuals of associations or species which present differences of structure, color, etc., as defined by naturalists; that the individuals of any such series or species are incapable of reproducing with those of any other species, with some exceptions; and that in the latter cases the offspring are usually entirely infertile.

The hypothesis of Cuvier assumes that each species was created by Divine power as we now find it at some definite point of geologic time. The paleontologist holding this view sees, in accordance therewith, a succession of creations and destructions marking the history of life on our planet from its commencement.

The development hypothesis states that all existing species have been derived from species of preëxistent geological periods, as offspring or by direct descent ; that there have been no total destructions of life in past time, but only a transfer of it from place to place, owing to changes of circumstance ; that the types of structure become simpler ànd more similar to each other as we trace them from later to earlier periods; and that finally we reach the simplest forms consistent with one or several original parent types of the great divisions into which living beings naturally fall.

It is evident, therefore, that the hypothesis does not include change of species by hybridization, nor allow the descent of living species from any other *living* species : both these propositions are errors of misapprehension or misrepresentation.

In order to understand the history of creation of a complex being, it is necessary to analyze it and ascertain of what it consists. In analyzing the construction of an animal or plant we readily arrange its characters into those which it possesses in common with other animals or plants, and those in which it resembles none other : the latter are its *individual* characters, constituting its individuality. Next we find a large body of characters, generally of a very obvious kind, which it possesses in common with a generally large number of individuals, which, taken collectively, all men are accustomed to call a species ; these characters we consequently name *specific*. Thirdly, we find characters, generally in parts of the body which are of importance in the activities of the animal, or which lie in near relation to its mechanical construction in details, which are shared by a still larger number of individuals than those which were similar in specific characters. In other words, it is common to a large number of species. This kind of character we call *generic*, and the grouping it indicates is a genus.

Farther analysis brings to light characters of organism which are common to a still greater number of individuals; this we call a *family* character. Those which are common to still more numerous individuals are the *ordinal:* they are usually found in parts of the structure which have the closest connection with the whole life-history of the being. Finally, the individuals composing many orders will be found identical in some important character of the systems by which ordinary life is maintained, as in the nervous and circulatory: the divisions thus outlined are called *classes*.

By this process of analysis we reach in our animal or plant those peculiarities which are common to the whole animal or vegetable kingdom, and then we have exhausted the structure so completely that we have nothing remaining to take into account beyond the cell-structure or homogeneous protoplasm by which we know that it is organic, and not a mineral.

The history of the origin of a type, as species, genus, order, etc., is simply the history of the origin of the structure or structures which define those groups respectively. It is nothing more nor less than this, whether a man or an insect be the object of investigation.

EVIDENCES OF DERIVATION.

α. *Of Specific Characters.*

The evidences of derivation of species from species, within the limits of the genus, are abundant and conclusive. In the first place, the rule which naturalists observe in defining species is a clear consequence of such a state of things. It is not amount and degree of difference that determine the definition of species from species, but it is the *permanency* of the characters in all cases and under all circumstances. Many species of the systems include varieties and extremes of form, etc., which, were they at all times distinct, and not connected by intermediate forms, would be estimated as species by the same and other writers, as can be easily seen by reference to their works.

Thus, species are either "restricted" or "protean," the latter embracing many, the former few variations; and the varieties included by the protean species are often as different from each other in their typical forms as are the "restricted" species. As an example, the species *Homo sapiens* (man) will suffice. His primary varieties are as distinct as the species of many well known genera, but cannot be defined, owing to the ex-istence of innumerable intermediate forms between them.

As to the common origin of such "varieties" of the protean species, naturalists never had any doubt, yet when it comes to the restricted "species," the anti-de-velopmentalist denies it *in toto*. Thus the varieties of most of the domesticated animals are some of them known—others held with great probability to have had a common origin. Varieties of plumage in fowls and canaries are of every-day occurrence, and are produced under our eyes. The cart-horse and racer, the Shet-land pony and the Norman, are without doubt derived from the same parentage. The varieties of pigeons and ducks are of the same kind, but not every one is aware of the extent and amount of such variations. The varieties in many characters seen in hogs and cattle, especially when examples from distant countries are compared, are very striking, and are confessedly equal in degree to those found to *define* species in a state of nature : here, however, they are not *definitive*.

It is easy to see that all that is necessary to produce in the mind of the anti-developmentalist the illusion of distinct origin by creation of many of these forms, would be to destroy a number of the intermediate conditions of specific form and structure, and thus to leave remaining definable groups of individuals, and therefore "species."

That such destructions and extinctions have been going on ever since the existence of life on the globe is well known. That it should affect intermediate forms, such as bind together the types of a protean species as well as restricted species, is equally certain. That its result has been to produce *definable* species cannot be denied, especially in consideration of the following facts : Protean species nearly always have a wide geographical distribution. They exist under more varied circumstances than do individuals of a more restricted species. The subordinate variations of the protean species are generally, like the restricted species, confined to distinct subdivisions of the geographical area which the whole occupies. As in geological time changes of level have separated areas once continuous by bodies of water or high mountain ranges, so have vast numbers of individuals occupying such areas been destroyed. Important alterations of temperature, or great changes in abundance or character of vegetable life over given areas, would produce the same result.

This part of the subject might be prolonged, were it

necessary, but it has been ably discussed by Darwin. The *rationale* of the "origin of species" as stated by him may be examined a few pages farther on.

✒

β. *Of the Characters of Higher Groups.*

a. Relations of Structures. The evidences of derivative origin of the structures defining the groups called genera, and all those of higher grade, are of a very different character from those discussed in relation to specific characters ; they are more difficult of observation and explanation.

Firstly: It would appear to be supposed by many that the creation of organic types was an irregular and capricious process, variously pursued by its Author as regards time and place, and without definite final aim ; and this notwithstanding the wonderful evidences we possess, in the facts of astronomy, chemistry, sound, etc., of His adhesion to harmonious and symmetrical sequences in His modes and plans.

Such regularity of plan is found to exist in the relations of the great divisions of the animal and vegetable kingdoms as at present existing on the earth. Thus, with animals we have a great class of species which consists of nothing more than masses or cells of protoplasmic matter, without distinct organs ; or the Protozoa. We have then the Cœlenterata (example, corals,) where the organism is composed of many cells arranged in distinct parts, but where a single very simple system of organs, forming the only internal cavity of the body, does the work of the many systems of the more complex animals. Next, the Echinodermata (such as starfish) present us with a body containing distinct systems

of organs enclosed in a visceral cavity, including a rudimental nervous system in the form of a ring.   In the
Molluscs to this condition is added additional complication, including extensions of the nervous system from
the ring as a starting-point, and a special organ for a
heart.   In the Articulates (crabs, insects,) we have like
complications, and a long distinct nervous axis on the
lower surface of the body.   The last branch or division
of animals is considered to be higher, because all the
systems of life organs are most complex or specialized.
The nervous ring is almost obliterated by a great en-
largement of its usual ganglia, thus become a brain,
which is succeeded by a long axis on the upper side
of the body.   This and other points define the Vertebrata.

Plans of structure, independent of the simplicity or
perfection of the special arrangement or structure of
organs, also define these great groups.   Thus the Protozoa present a spiral, the Cœlenterata a radiate, the
Echinodermata a bilateral radiate plan.   The Articulates are a series of external rings, each in one or more
respects repeating the others.   The Molluscs are a sac,
while a ring above a ring, joined together by a solid
center-piece, represents the plan of each of the many
segments of the Vertebrates which give the members of
that branch their form.

These bulwarks of distinction of animal types are
entered into here simply because they are the most in-
violable and radical of those with which we have to
deal, and to give the anti-developmentalist the best foot-
hold for his position.   I will only allude to the relations
of their points of approach, as these are affected by
considerations afterward introduced.

The Vertebrates approach the Molluscs at the lowest extreme of the former and higher of the latter. The lamprey eels of the one possess several characters in common with the cuttle-fish or squids of the latter. The amphioxus is called the lowest Vertebrate, and though it is nothing else, the definition of the division must be altered to receive it ; it has no brain !

The lowest forms of the Molluscs and Articulates are scarcely distinguishable from each other, so far as adhesion to the "plan" is concerned, and some of the latter division are very near certain Echinodermata. As we approach the boundary-lines of the two lowest divisions, the approaches become equally close, and the boundaries very obscure.

More instructive is the evidence of the relation of the subordinate classes of any one of these divisions. The conditions of those organs or parts which define classes exhibit a regular relation, commencing with simplicity and ending with complication; first associated with weak exhibitions of the highest functions of the nervous system—at the last displaying the most exalted traits found in the series.

For example : In the classes of Vertebrates we find the lowest nervous system presents great simplicity— the brain cannot be recognized ; next (in lampreys), the end of the nervous axis is subdivided, but scarcely according to the complex type that follows. In fishes the cerebellum and cerebral hemispheres are minute, and the intermediate or optic lobes very large : in the reptiles the cerebral hemispheres exceed the optic lobes, while the cerebellum is smaller. In birds the cerebellum becomes complex and the cerebrum greatly increases. In mammals the cerebellum increases in complexity or number of parts, the optic lobes diminish, while the cerebral hemispheres become wonderfully complex and enlarged, bringing us to the highest development, in man.

The history of the circulatory system in the Vertebrates is the same.* First, a heart with one chamber, then one with two divisions : three divisions belong to a large series, and the highest possess four. The origins of the great artery of the body, the aorta, are first five on each side : they lose one in the succeeding class in the ascending scale, and one in each succeeding class or order, till the Mammalia, including man, present us with but one on one side.

From an infinitude of such considerations as the above, we derive the certainty that the general arrangement of the various groups of the organic world is in scales, the subordinate within the more comprehensive divisions. The identification of all the parts in such a complexity of organism as the highest animals present, is a matter requiring much care and attention, and constitutes the study of homologies. Its pursuit has resulted in the demonstration that every individual of every species of a given branch of the animal kingdom is composed of elements common to all, and that the differences which are so radical in the higher groups are but the modifications of the same elemental parts, representing completeness or incompleteness, obliteration or subdivision. Of the former character are rudimental organs, of which almost every species possesses an example in some part of its structure.

But we have other and still more satisfactory evidence of the meaning of these relations. By the study of embryology we can prove most indubitably that the simple and less complex are inferior to the more complex.

* See a homological system of the circulatory system in the author's Origin of Genera, p. 22.

Selecting the Vertebrates again as an example, the highest form of mammal—*e. g.*, man—presents in his earliest stages of embryonic growth a skeleton of cartilage, like that of the lamprey: he also possesses five origins of the aorta and five slits on the neck, both which characters belong to the lamprey and the shark.   If the whole number of these parts does not coëxist in the embryonic man, we find in embryos of lower forms more nearly related to the lamprey that they do.   Later in the life of the mammal but four aortic origins are found, which arrangement, with the heart now divided into two chambers, from a beginning as a simple tube, is characteristic of the class of Vertebrates next in order—the bony fishes.   The optic lobes of the human brain have also at this time a great predominance in size—a character above stated to be that of the same class.   With advancing development the infant mammal follows the scale already pointed out.   Three chambers of the heart and three aortic origins follow, presenting the condition permanent in the batrachia ; and two origins, with enlarged cerebral hemispheres of the brain, resemble the reptilian condition.   Four heart-chambers, and one aortic root on each side, with slight development of the cerebellum, follow all characters defining the crocodiles, and immediately precede the special conditions defining the mammals.   These are, the single aorta root from one side, and the full development of the cerebellum : later comes that of the cerebrum also in its higher mammalian and human traits.

Thus we see the order already pointed out to be true, and to be an ascending one.   This is the more evident as each type or class passes through the conditions of those below it, as did the mammal ; each scale being

shorter as its highest terminus is lower.   Thus the crocodile passes through the stage of the lamprey, the fish,
the batrachian and the reptile proper.

b. In Time. We have thus a scale of relations of
existing forms of animals and plants of a remarkable
kind, and such as to stimulate greatly our inquiries as
to its significance.   When we turn to the remains of the
past creation preserved to us in the deposits continued
throughout geologic time, we are not disappointed, for
great light is at once thrown upon the subject.

We find, in brief, that the lowest division of the animal kingdom appeared first, and long before any type
of a higher character was created.   The Protozoön,
Eozoön, is the earliest of animals in geologic time, and
represents the lowest type of animal life now existing.
We learn also that the highest branch appeared last.
No remains of Vertebrates have been found below the
lower Devonian period, or not until the Echinoderms
and Molluscs had reached a great preëminence.   It is
difficult to be sure whether the Protozoa had a greater
numerical extent in the earliest periods than now, but
there can be no doubt that the Cœlenterata (corals) and
Echinoderms (crinoids) greatly exceeded their present
bounds, in Paleozoic time, so that those at present existing are but a feeble remnant.   If we examine the
subdivisions known as classes, evidence of the nature
of the succession of creation is still more conclusive.
The most polyp-like of the Molluscs (brachiopoda) constituted the great mass of its representatives during
Paleozic time.   Among Vertebrates the fishes appear
first, and had their greatest development in size and
numbers during the earliest periods of the existence of
the division.   Batrachia were much the largest and

most important of land animals during the Carboniferous period, while the higher Vertebrates were unknown. The later Mesozoic periods saw the reign of
reptiles, whose position in structural development has
been already stated. Finally, the most perfect, the
mammal, came upon the scene, and in his humblest
representatives. In Tertiary times mammalia sup-
planted the reptiles entirely, and the unspiritual mammals now yield to man, the only one of his class in
whom the Divine image appears.

Thus the structural relations, the embryonic characters, and the successive appearance in time of animals
coincide. The same is very probably true of plants.

That the existing state of the geological record of
organic types should be regarded as anything but a
fragment is, from our stand-point, quite preposterous.
And more, it may be assumed with safety that when
completed it will furnish us with a series of regular suc-
cessions, with but slight and regular interruptions, if
any, from the species which represented the simplest
beginnings of life at the dawn of creation, to those
which have displayed complication and power in later
or in the present period.

For the labors of the paleontologist are daily bringing to light structures intermediate between those never
before so connected, and thus creating lines of succession where before were only interruptions. Many such
instances might be adduced: two may be selected as
examples from American paleontology ;* *i. e.*, the near

---

* Professor Huxley, in the last anniversary lecture before the
Geological Society of London, recalls his opinion, enunciated in
1862, that "the positively-ascertained truths of Paleontology"
negative "the doctrines of progressive modification, which suppose

approach to birds made by the reptiles Lælaps and Megadactylus ; and the combination of characters of the sub-orders of Cryptodire and Pleurodire Tortoises in the Adocus of New Jersey.

---

that modification to have taken place by a necessary progress from more to less embryonic forms, from more to less generalized types, within the limits of the period represented . by . the fossiliferous rocks ; that it shows no evidence of such modification ; and as to the nature of that modification, it yields no evidence whatsoever that the earlier members of any long-continued group were more generalized in structure than the later ones."

Respecting this position, he says : " Thus far I have endeavored to expand and enforce by fresh arguments, but not to modify in any important respect, the ideas submitted to you on a former occasion. But when I come to the propositions respecting progressive modi-fication, it appears to me, with the help of the new light which has broken from various quarters, that there is much ground for soften-ing the somewhat Brutus-like severity with which I have dealt with a doctrine for the truth of which I should have been glad enough to be able to find a good foundation in 1862. So far indeed as the Invertebrata and the lower Vertebrata are concerned, the facts, and the conclusions which are to be drawn from them, appear to me to remain what they were. For anything that as yet appears to the con-trary, the earliest known marsupials may have been as highly organ-ized as their living congeners ; the Permian lizards show no signs of inferiority to those of the present day ; the labyrinthodonts can-not be placed below the living salamander and triton ; the Devonian ganoids are closely related to polypterus and lepidosiren."

To this it may be replied : 1. The scale of progression of the Vertebrata is measured by the conditions of the circulatory system, and in some measure by the nervous, and not by the osseous : tested by this scale, there has been successional complication of structure among Vertebrata in time. 2. The question with the evolutionist is, not what types have persisted to the present day, but the order in which types appeared in time. 3. The Marsupials, Permian saurians, labyrinthodonts and Devonian ganoids are re-markably generalized groups, and predecessors of types widely separated in the present period. 4. Professor Huxley adduces

We had no more reason to look for intermediate or. connecting forms between such types as these, than between any others of similar degree of remove from each other with which we are acquainted. And inasmuch as almost all groups, as genera, orders, etc., which are held to be distinct, but adjacent, present certain points of approximation to each other, the almost daily discovery of intermediate forms gives us confidence to believe that the pointings in other cases will also be realized.

γ. Of Transitions.

The preceding statements were necessary to the comprehension of the supposed mode of metamorphosis or development of the various types of living beings, or, in other words, of the single structural features which define them. . . . As it is evident that the more comprehensive groups, or those of highest rank, have

---

many such examples among the mammalian subdivisions in the remaining portion of his lecture. 5. Two alternatives are yet open in the explanation of the process of evolution : since generalized types, which combine the characters of higher and lower groups of later periods, must thus be superior to the lower, the lower must (first) be descended from such a generalized form by degradation ; or (second) not descended from it at all, but from some lower contemporaneous type by advance ; the higher only of the two being derived from the first-mentioned. The last I suspect to be a true explanation, as it is in accordance with the homologous groups. This law will shorten the demands of paleontologists for time, since, instead of deriving all reptilia, batrachia, etc., from common origins, it points to the derivation of higher reptilia of a higher order from higher reptilia of a lower order, lower reptilia of the first from lower reptilia of the second ; finally, the several groups of the lowest or most generalized order of reptilia from a parallel series of the class below, or batrachia.

had their origin in remote ages, cases of transition from
one to the other by change of character cannot be witnessed at the present day. We therefore look to the
most nearly related divisions, or those of the lowest
rank, for evidence of such change.

It is necessary to premise that embryology teaches
that all the species of a given branch of the animal kingdom (*e. g.*, Vertebrate, Mollusc, etc.) are quite identical
in structural character at their first appearance on the
germinal layer of the yolk of the parent egg. It shows
that the character of the respective groups of high rank
appear first, then those of less grade, and last of all
those structures which distinguish them as genera. But
among the earliest characters which appear are those of
the species, and some of those of the individual.

We find the characters of different *genera* to bear the
same relation to each other that we have already seen
in the case of those definitive of orders, etc. In a natural assemblage of related genera we discover that some
are defined by characters found only in the embryonic
stages of others; while a second will present a permanent condition of its definitive part, which marks a more
advanced stage of that highest. In this manner many
stages of the highest genus appear to be represented by
permanent genera in all natural groups. Generally,
however, this resemblance does not involve an entire
identity, there being some other immaturities found in
the highest genus at the time it presents the character
preserved in permanency by the lower, which the lower
loses. Thus (to use a very coarse example) a frog at
one stage of growth has four legs and a tail: the salamander always preserves four legs and a tail, thus resembling the young frog. The latter is, however, not a

salamander at that time, because, among other things, the skeleton is represented by cartilage only, and the salamander's is ossified. This relation is therefore an imitation only, and is called *inexact parallelism*.

As we compare nearer and nearer relations—*i. e.*, the genera which present fewest points of difference—we find the differences between undeveloped stages of the higher and permanent conditions of the lower to grow fewer and fewer, until we find numerous instances where the lower genus is exactly the same as the undeveloped stage of the higher. This relation is called that of *exact parallelism*.

It must now be remembered that the permanence of a character is what gives it its value in defining genus, order, etc., in the eyes of the systematist. So long as the condition is permanent no transition can be seen: there is therefore no development. If the condition is transitional, it defines nothing, and nothing is developed ; at least, so says the anti-developmentalist. It is the old story of the settler and the Indian: " Will you take owl and I take turkey, or I take turkey and you owl ?"

If we find a relation of *exact parallelism* to exist between two sets of species in the condition of a certain organ, and the difference so expressed the only one which distinguishes them as sets from each other—if that condition is always the same in each set—we call them two genera : if in any species the condition is variable at maturity, or sometimes the undeveloped condition of the part is persistent and sometimes transitory, the sets characterized by this difference must be united by the systematist, and the whole is called a single genus.

, We know numerous cases where different individuals of the same species present this relation of. *exact parallelism* to each other ; and as we ascribe common origin to the individuals of a species, we are assured that the condition of the inferior individual is, in this case, simply one of repressed growth, or a failure to fulfill the course accomplished by the highest. Thus, certain species of the salamandrine genus amblystoma undergo a metamorphosis involving several parts of the osseous and circulatory systems, etc., while half grown ; others delay it till fully grown ; one or two species remain indifferently unchanged or changed, and breed in either condition, while another species breeds unchanged, and has never been known to complete a metamorphosis.

The nature of the relation of *exact parallelism* is thus explained to be that of checked or advanced growth of individuals having a common origin. The relation of *inexact parallelism* is readily explained as follows : With a case of *exact parallelism* in the mind, let the repression producing the character of the lower, parallelize the latter with a stage of the former in which a second part is not quite mature: we will have a slight want of correspondence between the two. The lower will be immature in but one point, the incompleteness of the higher being seen in two points. If we suppose the immaturity to consist in a repression at a still earlier point in the history of the higher, the latter will be undeveloped in other points also : thus, the spike-horned deer of South America have the horn of the second year of the North American genus. They would be generically identical with that stage of the latter, were it not that these still possess their milk dentition at two years of age. In the same way the nature of the parallelisms seen

in higher groups, as orders, etc., may be accounted for.
The theory of homologous groups furnishes important evidence in favor of derivation. ' Many orders of
animals (probably all, when we come to know them) are
divisible into two or more sections, which I have called
homologous. These are series of genera or families,
which differ from each other by some marked character,
but whose contained genera or families differ from each
other in the same points of detail, and in fact correspond exactly. So striking is this correspondence that
were it not for the general and common character separating the homologous series, they would be regarded as
the same, each to each. Now it is remarkable that
where studied the difference common to all the terms of
two homologous groups is found to be one of *inexact
parallelism,* which has been shown above to be evidence
of descent. Homologous groups always occupy different geographical areas on the earth's surface, and their
relation is precisely that which holds between successive groups of life in the periods of geologic time.

In a word, we learn from this source that distinct geologic epochs coëxist at the same time on the earth. I
have been forced to this conclusion* by a study of the
structure of terrestrial life, and it has been remarkably
confirmed by the results of recent deep-sea dredgings
made by the United States Coast Survey in the Gulf
Stream, and by the British naturalists in the North Atlantic. These have brought to light types of Tertiary
life, and of even the still more ancient Cretaceous periods, living at the present day. That this discovery
invalidates in any wise the conclusions of geology re-

* *Origin of Genera,* pages 70, 77, 79.

specting lapse of time is an unwarranted assumption that some are forward to make. If it changes the views of some respecting the parallelism or coëxistence of faunæ in different regions of the earth, it is only the anti-developmentalists whose position must be changed.

For, if we find distinct geologic faunæ, or epochs defined by faunæ, coëxisting during the present period, and fading or emerging into one another as they do at their geographical boundaries, it is proof positive that the geologic epochs and periods of past ages had in like manner no trenchant boundaries, but also passed the one into the other. The assumption that the apparent interruptions are the result of transfer of life rather than destruction, or of want of opportunities of preservation, is no doubt the true one.

δ. Rationale of Development.

a. In Characters of Higher Groups. It is evident in the case of the species in which there is an irregularity in the time of completion of metamorphosis that some individuals traverse a longer developmental line than those who remain more or less incomplete. As both accomplish growth in the same length of time, it is obvious that it proceeds with greater rapidity in one sense in that which accomplishes most : its growth is said to be accelerated. This phenomenon is especially common among insects, where the females of perfect males are sometimes larvæ or nearly so, or pupæ, or lack wings or some character of final development. Quite as frequently, some males assume characters in advance of others, sometimes in connection with a peculiar geographical range.

· In cases of *exact parallelism* we reasonably suppose the cause to be the same, since the conditions are identical, as has been shown ; that is, the higher conditions have been produced by a crowding back of the earlier characters and an acceleration of growth, so that a given succession in order of advance has extended over a longer range of growth than its predecessor in the same allotted time. That allotted time is the period before maturity and reproduction, and it is evident that as fast as modifications or characters should be assumed sufficiently in advance of that period, so certainly would they be conferred upon the offspring by reproduction. The *acceleration* in the assumption of a character, progressing more rapidly than the same in another character, must soon produce, in a type whose stages were once the exact parallel of a permanent lower form, the condition of *inexact parallelism.* As all the more comprehensive groups present this relation to each other, we are compelled to believe that *acceleration* has been the principle of their successive evolution during the long ages of geologic time.

Each type has, however, its day of supremacy and perfection of organism, and a retrogression in these respects has succeeded. This has no doubt followed a law the reverse of acceleration, which has been called *retardation.* By the increasing slowness of the growth of the individuals of a genus, and later and later assumption of the characters of the latter, they would be successively lost.

To what power shall we ascribe this acceleration, by which the first beginnings of structure have accumulated to themselves through the long geologic ages complication and power, till from the germ that was scarcely born into a sand-lance, a human being climbed the complete scale, and stood easily the chief of the whole ?

In the cases of species, where some individuals develop farther than others, we say the former possess more growth-force, or "vigor," than the latter. We may therefore say that higher types of structure possess more "vigor" than the lower. This, however, we do not know to be true, nor can we readily find means to demonstrate it.

The food which is taken by an adult animal is either assimilated, to be consumed in immediate activity of some kind, or stored for future use, and the excess is rejected from the body. We have no reason to suppose that the same kind of material could be made to subserve the production of life-force by any other means than that furnished by a living animal organism. The material from which this organism is constructed is derived first from the parent, and afterward from the food, etc., assimilated by the individual itself so long as growth continues. As it is the activity of assimilation directed to a special end during this latter period which we suppose to be increased in accelerated development, the acceleration is evidently not brought about by increased facilities for obtaining the means of life which the same individual possesses as an adult. That it is not in consequence of such increased facilities possessed by its parents over those of the type preceding it, seems equally improbable when we consider that the characters in which the parent's advance has appeared are rarely of a nature to increase those facilities.

The nearest approach to an explanation that can be offered appears to be somewhat in the following direction:

There is every reason to believe that the character of the atmosphere has gradually changed during geologic time, and that various constituents of the mixture have been successively removed from it, and been stored in the solid material of the earth's crust in a state of combination. Geological chemistry has shown that the cooling of the earth has been accompanied by the precipitation of many substances only gaseous at high temperatures. Hydrochloric and sulphuric acids have been transferred to mineral deposits or aqueous solutions. The removal of carbonic acid gas and the vapor of water has been a process of much slower progress, and after the expiration of all the ages a proportion of both yet remains. Evidence of the abundance of the former in the earliest periods is seen in the vast deposits of limestone rock ; later, in the prodigious quantities of shells which have been elaborated from the same in solution. Proof of its abundance in the atmosphere in later periods is seen in the extensive deposits of coal of the Carboniferous, Triassic and Jurassic periods. If the most luxuriant vegetation of the present day takes but fifty tons of carbon from the atmosphere in a century, per acre, thus producing a layer over that extent of less than a third of an inch in thickness, what amount of carbon must be abstracted in order to produce strata of thirty-five feet in depth ? No doubt it occupied a long period, but the atmosphere, thus deprived of a large proportion of carbonic acid, would in subsequent periods undoubtedly possess an improved capacity for the support of animal life.

The successively higher degree of oxidization of the blood in the organs designed for that function, whether performing it in water or air, would certainly accelerate

2

the performances of all the vital functions, and among others that of growth. Thus it may be that *acceleration* can be accounted for, and the process of the development of the orders and sundry lesser groups of the Vertebrate kingdom indicated ; for, as already pointed out, the definitions of such are radically placed in the different structures of the organs which aerate the blood and distribute it to its various destinations.

But the great question, What determined the direction of this acceleration? remains unanswered. One cannot understand why more highly-oxidized blood should hasten the growth of partition of the ventricle of the heart in the serpent, the more perfectly to separate the aerated from the impure fluid ; nor can we see why a more perfectly-constructed circulatory system, sending purer blood to the brain, should direct accelerated growth to the cerebellum or cerebral hemispheres in the crocodile.

b. In Characters of the Specific Kind. Some of the characters usually placed in the specific category have been shown to be the same in kind as those of higher categories. The majority are, however, of a different kind, and have been discussed several pages back.

The cause of the origin of these characters is shrouded in as much mystery as that of those which have occupied the pages immediately preceding. As in that case, we have to assume, as Darwin has done, a tendency in Nature to their production. This is what he terms "the principle of variation." Against an unlimited variation the great law of heredity or atavism has ever been opposed, as a conservator and multiplier of type. This principle is exemplified in the fact that like produces like—that children are like their parents, frequently even in minutiæ. It may be compared to habit in metaphysical matters, or to that singular love of time or rhythm seen in man and lower animals, in both of which the tendency is to repeat in continual cycles a motion or state of the mind or sense.

Further, but a proportion of the lines of variation is supposed to have been perpetuated, and the extinction of intermediate forms, as already stated, has left isolated groups or species.

The effective cause of these extinctions is stated by Darwin to have been a "natural selection"—a proposition which distinguishes his theory from other development hypotheses, and which is stated in brief by the expression, "the preservation of the fittest." Its meaning is this: that those characters appearing as results of this spontaneous variation which are little adapted to the conflict for subsistence, with the nature of the supply, or with rivals in its pursuit, dwindle and are sooner or later extirpated; while those which are adapted to their surroundings, and favored in the struggle for means of life and increase, predominate, and ultimately become the centers of new variation. "I am convinced," says Darwin, "that natural selection has been the main, but not exclusive, means of modification."

That it has been to a large extent the means of preservation of those structures known as specific, must, I think, be admitted. They are related to their peculiar surroundings very closely, and are therefore more likely to exist under their influence. Thus, if a given genus extends its range over a continent, it is usually found to be represented by peculiar species—one in a maritime division, another in the desert, others in the forest, in the swamp or the elevated areas of the region. The wonderful interdependence shown by Darwin to exist between insects and plants in the fertilization of the latter, or between animals and their food-plants, would almost induce one to believe that it were the true expression of the whole law of development.

But the following are serious objections to its universal application :

First: The characters of the higher groups, from genera up, are rarely of a character to fit their possessors especially for surrounding circumstances ; that is, the differences which separate genus from genus, order from order, etc., in the ascending scale of each, do not seem to present a superior adaptation to surrounding circumstances in the higher genus to that seen in the lower genus, etc. Hence, superior adaptation could scarcely have caused their selection above other forms not existing. Or, in other words, the different structures which indicate successional relation, or which measure the steps of progress, seem to be equally well fitted for the same surroundings.

Second : The higher groups, as orders, classes, etc., have been in each geologic period alike distributed over the whole earth, under all the varied circumstances offered by climate and food. Their characters do not seem to have been modified in reference to these. Species, and often genera, are, on the other hand, eminently restricted according to climate, and consequently vegetable and animal food.

The law of development which we seek is indeed not that which preserves the higher forms and rejects the lower after their creation, but that which explains why higher forms were created at all. Why in the results of a creation we see any relation of higher and lower, and not rather a world of distinct types, each perfectly adapted to its situation, but none properly higher than another in an ascending scale, is the primary question. Given the principle of advance, then natural selection has no doubt modified the details; but in the successive advances we can scarcely believe such a principle to be influential. *We look rather upon a progress as the result of the expenditure of some force fore-arranged for that end.*

It may become, then, a question whether in characters of high grade the habit or use is not rather the result of the acquisition of the structure than the structure the result of the encouragement offered to its assumed beginnings by use, or by liberal nutrition derived from the increasingly superior advantages it offers.

8. *The Physical Origin of Man.*

If the hypothesis here maintained be true, man is the. descendant of some preëxistent generic type, the which, if it were now living, we would probably call an ape.

Man and the chimpanzee were in Linnæus' system only two species of the same genus, but a truer anatomy places them in separate genera and distinct families. There is no doubt, however, that Cuvier went much too far when he proposed to consider Homo as the representative of an order distinct from the quadrumana, under the name of bimana. The structural differences will not bear any such interpretation, and have not the same value as those distinguishing the orders of mammalia; as, for instance, between carnivora and bats, or the cloven-footed animals and the rodents, or rodents and edentates. The differences between man and the chimpanzee are, as Huxley well puts it, much less than those between the chimpanzee and lower quadrumana, as lemurs, etc. In fact, man is the type of a family, Hominidæ, of the order Quadrumana, as indicated by the characters of the dentition, extremities, brain, etc. The reader who may have any doubts on this score may read the dissections of Geoffroy St. Hilaire, made in 1856, before the issue of Darwin's *Origin of Species*. He informs us that the brain of man is nearer in structure to that of the orang than the orang's is to that of the South American howler, and that the orang and howler are more nearly related in this regard than are the howler and the marmoset.

The modifications presented by man have, then, resulted from an acceleration in development in some respects, and retardation perhaps in others. But until the *combination* now characteristic of the genus Homo was attained the being could not properly be called man.

And here it must be observed that as an organic type is characterized by the coëxistence of a number of peculiarities which have been developed independently of each other, its distinctive features and striking functions are not exhibited until that coëxistence is attained which is necessary for these ends.

Hence, the characters of the human genus were probably developed successively ; but few of the indications of human superiority appeared until the combination was accomplished. Let the opposable thumb be first perfected, but of what use would it be in human affairs without a mind to direct? And of what use a mind without speech to unlock it? And speech could not be possible though all the muscles of the larynx but one were developed, or but a slight abnormal convexity in one pair of cartilages remained.

It would be an objection of little weight could it be truly urged that there have as yet no remains of ape-like men been discovered, for we have frequently been called upon in the course of paleontological discovery to bridge greater: gaps than this, and greater remain, which we expect to fill. But we *have* apelike characters exhibited by more than one race of men yet existing.

But the remains of that being which is supposed to have been the progenitor of man may have been discovered a short time since in the cave of Naulette, Belgium, with the bones of the extinct rhinoceros and elephant.

We all admit the existence of higher and lower races, the latter being those which we now find to present greater or less approximations to the apes. The peculiar structural characters that belong to the negro in his most typical form are of that kind, however great may be the distance of his remove therefrom. The flattening of the nose and prolongation of the jaws constitute such a resemblance ; so are the deficiency of the calf of the leg, and the obliquity of the pelvis, which approaches more the horizontal position than it does in the Caucasian. The investigations made at Washington during the war with reference to the physical characteristics of the soldiers show that the arms of the negro are from one to two inches longer than those of the whites : another approximation to the ape. In fact, this race is a species of the genus Homo, as distinct in character from the Caucasian as those we are accustomed to recognize in other departments of the animal kingdom ; but he is not distinct by isolation, since intermediate forms between him and the other species can be abundantly found,

And here let it be particularly observed that two of the most prominent characters of the negro are those of immature stages of the Indo-European race in its characteristic types. The deficient calf is the character of infants at a very early stage ; but, what is more important, the flattened bridge of the nose and shortened nasal cartilages are universally immature conditions of the same parts in the Indo-European. Any one may convince himself of that by examining the physiognomies of infants. In some races—*e. g.*, the Slavic—this undeveloped character persists later than in some others. The Greek nose, with its elevated bridge, coincides not only with æsthetic beauty, but with developmental perfection.

This is, however, only "*inexact* parallelism," as the characters of the hair, etc., cannot be explained on this principle *among existing races*. The embryonic characters mentioned are probably a remnant of those characteristic of the primordial race or species.

But the man of Naulette, if he be not a monstrosity, in a still more distinct and apelike species. The chin, that marked character of other species of men, is totally wanting, and the dentition is quite approximate to the man-like apes, and different from that of modern men. The form is very massive, as in apes. That he was not abnormal is rendered probable by approximate characters seen in a jaw from the cave of Puy-sur-Aube, and less marked in the lowest races of Australia and New Caledonia.

As to the single or multiple origin of man, science as yet furnishes no answer. It is very probable that, in many cases, the species of one genus have descended from corresponding species of another by change of generic characters only. It is a remarkable fact that the orang possesses the peculiarly developed malar bones and the copper color characteristic of the Mongolian inhabitants of the regions in which this animal is found, while the gorilla exhibits the prognathic jaws and black hue of the African races near whom he dwells. This kind of geographical imitation is very common in the animal kingdom.

ζ. *The Mosaic Account.*

As some persons imagine that this hypothesis conflicts with the account of the creation of man given in Genesis, a comparison of some of the points involved is made below.

First: In Genesis i. 26, 27, we read, " And God said, Let us make man in our image, after our likeness," etc. " So God created man in his own image, in the image of God created he him ; male and female created he them." Those who believe that this "image" is a physical, material form, are not disposed to admit the entrance of anything ape-like into its constitution, for the ascription of any such appearance to the Creator would be impious and revolting. But we are told that "God is a Spirit," and Christ said to his disciples after his resurrection, " A spirit hath not flesh and bones, as ye see me have." Luke xxiv. 39. It will require little further argument to show that a mental and spiritual image is what is meant, as it is what truly exists. Man's conscience, intelligence and creative ingenuity show that he possesses an "image of God " within him, the possession of which is really necessary to his limited comprehension of God and of God's ways to man.

2*

Second : In Genesis ii. 7, the text reads, "And the Lord God formed man of the dust of the ground, and breathed into his nostrils the breath of life ; and man became a living soul." The fact that man is the result of the modification of an ape-like predecessor nowise conflicts with the above statement as to the materials of which his body is composed. Independently of origin, if the body of man be composed of dust, so must that of the ape be, since the composition of the two is identical. But the statement simply asserts that man was created of the same materials which compose the earth : their condition as "dust " depending merely on tempera· ture and subdivision. The declaration, "Dust thou art, and unto dust thou shalt return," must be taken in a similar sense, for we know that the decaying body is re-solved not only into its earthly constituents, but also into carbonic acid gas and water.

When God breathed into man's nostrils the breath of life, we are informed that he became, not a living body, but "a living soul." His descent from a preëxistent being involved the possession of a living body ; but when the Creator breathed into him we may suppose for the present that He infused into this body the immortal part, and at that moment man became a conscientious and responsible being.

II. Metaphysical Evolution.

It is infinitely improbable that a being endowed with such capacities for gradual progress as man has exhibited, should have been full fledged in accomplishments at the moment when he could first claim his high title, and abandon that of his simious ancestors. We are

therefore required to admit the growth of human intelligence from a primitive state of inactivity and absolute
ignorance ; including the development of one important
mode of its expression—speech ; as well as that of the
moral qualities, and of man's social system—the form in
which his ideas of morality were first displayed.

The expression "evolution of morality" need not
offend, for the question in regard to the *laws* of this
evolution is the really important part of the discussion,
and it is to the opposing views on this point that the
most serious interest attaches.

*   *   *   *   *   *

The two views of evolution already treated of, held
separately, are quite opposed to each other.   The first
(and generally received) lays stress on the influence of
external surroundings, as the stimulus to and guidance
of development: it is the counterpart of Darwin's principle called Natural Selection in material progress.
This might be called the *Conflict theory*.   The second
view recognizes the workings of a force whose nature
we do not know, whose exhibitions accord perfectly with
their external surroundings (or other exhibitions of itself), without being under their influence or more related to them, as effect to cause, than the notes of the
musical octave or the colors of the spectrum are to each
other.   This is the *Harmonic theory*.   In other words,
the first principle deduces perfection from struggle and
discord ; the second, from the coincident progress of
many parts, forming together a divine harmony com-

parable to music. That these principles are both true is rendered extremely probable by the actual phenomena of development, material and immaterial. In other words, struggle and discord ever await that which is not in the advance, and which fails to keep pace with the harmonious development of the whole.

All who have studied the phenomena of the creation believe that there exists in it a grand and noble harmony, such as was described to Job when he was told that " the morning stars sang together, and all the sons of God shouted for joy."

α. *Development of Intelligence.*

If the brain is the organ of mind, we may be surprised to find that the brain of the intelligent man scarcely differs in structure from that of the ape. Whence, then, the difference of power? Though no one will now deny that many of the Mammalia are capable of reasoning upon observed facts, yet how greatly the results of this capacity differ in number and importance from those achieved by human intelligence! Like water at the temperatures of 50° and 53°, where we perceive no difference in essential character, so between the brains of the lower and higher monkeys no difference of function or of intelligence is perceptible. But what a difference do the two degrees of temperature from 33° to 31° produce in water! In like manner the difference between the brain of the higher ape and that of man is accompanied by a difference in function and power, on which, man's earthly destiny depends. In development, as with the water so with the higher ape : some Rubicon has been crossed, some floodgate has been opened, which marks one of Nature's great transitions, such as have been called "Expression points" of progress.

What point of progress in such a history would account for this accession of the powers of the human intelligence? It has been answered, with considerable confidence, The power of speech. Let us picture man without speech. Each generation would learn nothing from its predecessors. Whatever originality or observation might yield to a man would die with him. Each intellectual life would begin where every other life began, and would end at a point only differing with its original capacity. Concert of action, by which man's power over the material world is maintained, would not exceed, if it equaled, that which is seen among the bees; and the material results of his labors would not extend beyond securing the means of life and the employment of the simplest modes of defence and attack.

The first men, therefore, are looked upon by the developmentalists as extremely embryonic in all that characterizes humanity, and they appeal to the facts of history in support of this view. If they do not derive much assistance from written history, evidence is found in the more enduring relics of human handiwork.

The opposing view is, that the races which present or have presented this condition of inferiority or savagery have reached it by a process of degradation from a higher state—as some believe, through moral delinquency. This position may be true in certain cases, which represent perhaps a condition of senility, but in general we believe that savagery was the condition of the first man, which has in some races continued to the present day.

β. *Evidence from Archæology.*

As the object of the present essay is not to examine fully into the evidences for the theories of evolution here stated, but rather to give a sketch of such theories and their connection, a few facts only will be noticed.

Improvement in the use of Materials. As is well known, the remains of human handiwork of the earliest periods consist of nothing but rude implements of stone and bone, useful only in procuring food and preparing it for use. Even when enterprise extended beyond the ordinary routine, it was restrained by the want of proper instruments. Knives and other cutting implements of flint still attest the skill of the early races of men from Java to the Cape of Good Hope, from Egypt to Ireland, and through North and South America. Hatchets, spear-heads and ornaments of serpentine, granite, silex, clay slates, and all other suitable rock materials, are found to have been used by the first men, to the exclusion of metals, in most of the regions of the earth.

Later, the probably accidental discovery of the superiority of some of the metals resulted in the substitution of them for stone as a material for cutting implements. Copper—the only metal which, while malleable, is hard enough to bear an imperfect edge—was used by succeeding races in the Old World and the New. Implements of this material are found scattered over extensive regions. So desirable, however, did the hardening of the material appear for the improvement of the cutting edge that combinations with other metals were sought for and discovered. The alloy with tin, forming bronze and brass, was discovered and used in Europe, while that with silver appears to have been most readily produced in America, and was consequently used by the Peruvians and other nations.

The discovery of the modes of reducing iron ores placed in the hands of man the best material for bringing to a shape convenient for his needs the raw material of the world. All improvements in this direction made since that time have been in the quality of iron itself, and not through the introduction of any new metal.

The prevalent phenomena of any given period are those which give it its character, and by which we distinguish it. But this fact does not exclude the coëxistence of other phenomena belonging to prior or subsequent stages. Thus, during the many stages of human progress there have been men more or less in advance of the general body, and their characteristics have given a peculiar stamp to the later and higher condition of the whole. It furnishes no objection to this view that we find, as might have been anticipated, the stone, bronze and iron periods overlaping one another, or men of an inferior culture supplanting in some cases a superior people. A case of this kind is seen in North America, where the existing " Indians," stone-men, have succeeded the mound-builders, copper-men. The successional relation of discoveries is all that it is necessary to prove, and this seems to be established.

The period at which the use of metallic implements was introduced is unknown, but Whitney says that the language of the Aryans, the ancestors of all the modern Indo-Europeans, indicates an acquaintance with such implements, though it is not certain whether those of iron are to be included. The dispersion of the daughter races, the Hindoos, the Pelasgi, Teutons, Celts, etc.,

could not, it is thought, have taken place later than
3000 B. C.—a date seven hundred years prior, to that assigned by the old chronology to the Deluge. Those
races coëxisted with the Egyptian and Chinese nations,
already civilized, and as distinct from each other in
feature as they are now.

Improvement in Architecture. The earliest periods,
then, were characterized by the utmost simplicity of invention and construction. Later, the efforts for defence
from enemies and for architectural display, which have
always employed so much time and power, began to be
made. The megalithic period has left traces over much
of the earth. The great masses of stone piled on each
other in the simplest form in Southern India, and the
circles of stones planted on end in England at Stone-
henge and Abury, and in Peru at Sillustani, are relics
of that period. More complex are the great Himyaritic
walls of Arabia, the works of the ancestors of the
Phœnicians in Asia Minor, and the titanic workman-
ship of the Pelasgi in Greece and Italy. In the iron
age we find granitic hills shaped or excavated into temples ; as, for example, everywhere in Southern India.
Near Madura the circumference of an acropolis-like hill
is cut into a series of statues in high relief, of sixty feet
in elevation. Easter Island, composed of two volcanic
cones, one thousand miles from the west coast of South
America, in the bosom of the Pacific, possesses several
colossi cut from the intrusive basalt, some in high relief
on the face of the rock, others in detached blocks removed by human art from their original positions and
brought nearer the sea-shore.

Finally, at a more advanced stage, the more ornate
and complex structures of Central America, of Cam-

bodia, Nineveh and Egypt, represent the period of greatest display of architectural expenditure. The same amount of human force has perhaps never been expended in this direction since, though higher conceptions of beauty have been developed in architecture with increasing intellectuality.

Man has passed through the block-and-brick building period of his boyhood, and should rise to higher conceptions of what is the true disposition of power for "him who builds for aye," and learn that "spectacle" is often the unwilling friend of progress.

No traces of metallic implements have ever been found in the salt-mines of Armenia, the turquoise-quarries in Arabia, the cities of Central America or the excavations for mica in North Carolina, while the direct evidence points to the conclusion that in those places flint was exclusively used.

The simplest occupations, as requiring the least exercise of mind, are the pursuit of the chase and the tending of flocks and herds. Accordingly, we find our first parents engaged in these occupations. Cain, we are told, was, in addition, a tiller of the ground. Agriculture in its simplest forms requires but little more intelligence than the pursuits just mentioned, though no employment is capable of higher development. If we look at the savage nations at present occupying nearly half the land surface of the earth, we shall find many examples of the former industrial condition of our race preserved to the present day. Many of them had no knowledge of the use of metals until they obtained it from civilized men who visited them, while their pursuits were and are those of the chase, tending domestic animals, and rudimental agriculture.

γ. *The Development of Language.*

In this department the fact of development from the simple to the complex has been so satisfactorily demonstrated by philologists as scarcely to require notice here. The course of that development has been from monosyllabic to polysyllabic forms, and also in a process of differentiation, as derivative races were broken off from the original stock and scattered widely apart. The evidence is clear that simple words for distinct objects formed the bases of the primal languages, just as the ground, tree, sun and moon represent the character of the first words the infant lisps. In this department also the facts point to an infancy of the human race.

δ. *Development of the Fine Arts.*

If we look at representation by drawing or sculpture, we find that the efforts of the earliest races of which we have any knowledge were quite similar to those which the untaught hand of infancy traces on its slate or the savage depicts on the rocky faces of hills. The circle or triangle for the head and body, and straight lines for the limbs, have been preserved as the first attempts of the men of the stone period, as they are to this day the sole representations of the human form which the North American Indian places on his buffalo robe or mountain precipice. The stiff, barely-outlined form of the deer, the turtle, etc., are literally those of the infancy of civilized man.

The first attempts at sculpture were marred by the influence of modism. Thus the idols of Coban and Palenque, with human faces of some merit, are over-

loaded with absurd ornament, and deformed into fright-
ful asymmetry, in compliance with the demand of some
imperious mode. In later days we have the stiff, con-
ventionalized figures of the palaces of Nineveh and
the temples of Egypt, where the representation of form
has somewhat improved, but is too often distorted by
false fashion or imitation of some unnatural standard,
real or artistic. This is distinguished as the day of
archaic sculpture, which disappeared with the Etruscan
nation. So the drawings of the child, when he abandons the simple lines, are stiff and awkward, and but a
stage nearer true representation; and how often does
he repeat some peculiarity or absurdity of his own! So
much easier is it to copy than to conceive.

The introduction of the action and pose of life into
sculpture was not known before the early days of
Greece, and it was there that the art was brought to
perfection. When art rose from its mediæval slumber,
much the same succession of development may be discovered. First, the stiff figures, with straightened limbs
and cylindric drapery, found in the old Northern
churches—then the forms of life that now adorn the
porticoes and palaces of the cities of Germany.

ε. *Rationale of the Development of Intelligence.*

The history of material development shows that the
transition from stage to stage of development, experienced by the most perfect forms of animals and plants
in their growth from the primordial cell, is similar to the
succession of created beings which the geological
epochs produced. It also shows that the slow assumption of main characters in the line of succession in

early geological periods produced the condition of inferiority, while an increased rapidity of growth in later
days has resulted in an attainment of superiority. It is
not to be supposed that in "acceleration" the period
of growth is shortened: on the contrary, it continues
the same. Of two beings whose characters are assumed
at the same rate of succession, that with the quickest or
shortest growth is necessarily inferior. "Acceleration"
means a gradual increase of the rate of assumption of
successive characters in the same period of time. A
fixed rate of assumption of characters, with gradual increase in the length of the period of growth, would
produce the same result—viz., a longer developmental
scale and the attainment of an advanced position. The
first is in part the relation of sexes of a species ; the
last of genera, and of other types of creation. If from
an observed relation of many facts we derive a law, we
are permitted, when we see in another class of facts
similar relations, to suspect that a similar law has operated, differing only in its objects. We find a marked
resemblance between the facts of structural progress
in matter and the phenomena of intellectual and spiritual progress.

If the facts entering into the categories enumerated
in the preceding section bear us out, we conclude that
in the beginning of human history the progress of the
individual man was very slow, and that but little was
attained to ; that through the profitable direction of human energy, means were discovered from time to time
by which the process of individual development in all
metaphysical qualities has been accelerated ; and that
up to the present time the consequent advance of the
whole race has been at an increasing rate of progress,

This is in accordance with the general principle, that high development in intellectual things is accomplished by rapidity in traversing the preliminary stages of inferiority common to all, while low development signifies sluggishness in that progress, and a corresponding retention of inferiority.

How much meaning may we not see, from this standpoint, in the history of the intelligence of our little ones! First they crawl, they walk on all fours: when they first assume the erect position they are generally speechless, and utter only inarticulate sounds. When they run about, stones and dirt, the objects that first meet the eye, are the delight of their awakening powers, but these are all cast aside when the boy obtains his first jackknife. Soon, however, reading and writing open a new world to him ; and finally as a mature man he seizes the forces of nature, and steam and electricity do his bidding in the active pursuit of power for still better and higher ends.

So with the history of the species: first the quadrumane—then the speaking man, whose humble industry was, however, confined to the objects that came first to hand, this being the "stone age" of pre-historic time. When the use of metals was discovered, the range of industries expanded wonderfully, and the "iron age" saw many striking efforts of human power. With the introduction of letters it became possible to record events and experiences, and the spread of knowledge was thereby greatly increased, and the delays and mistakes of ignorance correspondingly diminished in the fields of the world's activity.

From the first we see in history a slow advance as knowledge gained by the accumulation of tradition and by improvements in habit based on experience; but how slow was this advance while the use of the metals was still unknown! The iron age brought with it not only new conveniences, but increased means of future progress; and here we have an acceleration in the rate of advance. With the introduction of letters this rate was increased many fold, and in the application of steam we have a change equal in utility to any that has preceded it, and adding more than any to the possibilities of future advance in many directions. By it power, knowledge and means of happiness were to be distributed among the many.

The uses to which human intelligence has successively applied the materials furnished by nature have been—First, subsistence and defence: second, the accumulation of power in the shape of a representative of that labor which the use of matter involves; in other words, the accumulation of wealth. The possession of this power involves new possibilities, for opportunity is offered for the special pursuits of knowledge and the assistance of the weak or undeveloped part of mankind in its struggles.

Thus, while the first men possessed the power of speech, and could advance a little in knowledge through the accumulation of the experiences of their predecessors, they possessed no means of accumulating the power of labor, no control over the activity of numbers —in other words, no wealth.

But the accumulation of knowledge finally brought this advance about. The extraction and utilization of the metals, especially iron, formed the most important step, since labor was thus facilitated and its productiveness increased in an incalculable degree. We have little evidence of the existence of a medium of exchange during the first or stone period, and no doubt barter was the only form of trade. Before the use of metals, shells and other objects were used: remains of money of baked clay have been found in Mexico. Finally, though in still ancient times, the possession of wealth in money gradually became possible and more common, and from that day to this avenues for reaching this stage in social progress has ever been opening.

But wealth merely indicates a stage of progress, since it is but a comparative term. All men could not become rich, for in that case all would be equally poor. But labor has a still higher goal; for, thirdly, as capital, it constructs and employs machinery, which does the work of many hands, and thus cheapens products, which is equivalent in effect to an accumulation of wealth to the consumer. And this increase of power may be used for the intellectual and spiritual advance of men, or otherwise, at the will of the men thus favored. Machinery places man in the position of a creator, operating on Nature through an increased number of " secondary causes."

Development of intelligence is seen, then, in the following directions: First, in the knowledge of facts, including science; second, in language; third, in the apprehension of beauty; and, as consequences of the first of these, the accumulation of power by development—First, of means of subsistence; and second, of mechanical invention.

Thus we have two terms to start with in estimating the beginning of human development in knowledge and power: First, the primary capacities of the human mind itself; second, a material world, whose infinitely varied components are so arranged as to yield results to the energies of that mind. For example, the transition points of vaporization and liquefaction are so placed as to be within the reach of man's agents; their weights are so fixed as to accord with the muscular or other forces which he is able to exert; and other living or-. ganizations are subject to his convenience and rule, and not, as in previous geological periods, entirely beyond his control. These two terms being given, it is maintained that the present situation of the most civilized men has been attained through the operation of a law of mutual action and reaction—a law whose results, seen at the present time, have depended on the acceleration or retardation of its rate of action; which rate has been regulated, according to the degree in which a third great term, viz., the law of moral or (what is the same thing) true religious development has been combined in the plan. What it is necessary to establish in order to prove the above hypothesis is—

I. That in each of the particulars above enumerated the development of the human species is similar to that of the individual from infancy to maturity.

II. That from a condition of subserviency to the laws of matter, man's intelligence enables him, by an accumulation of power, to become in a sense independent of those laws, and to increase greatly the rate of intellectual and spiritual progress.

III. That failure to accomplish a moral or spiritual development will again reduce him to a subserviency to the laws of matter.

This brings us to the subject of moral development. And here I may be allowed to suggest that the weight of the evidence is opposed to the philosophy, "falsely so called," of necessitarianism, which asserts that the first two terms alone were sufficient to work out man's salvation in this world and the next; and, on the other hand, to that anti-philosophy which asserts that all things in the progress of the human race, social and civil, are regulated by immediate Divine interposition instead of through instrumentalities. Hence the subject divides itself at once into two great departments— viz., that of the development of mind or intelligence, and that of the development of morality.

That these laws are distinct there can be no doubt, since in the individual man one of them may produce results without the aid of the other. Yet it can be shown that each is the most invaluable aid and stimulant to the other, and most favorable to the rapid advance of the mind in either direction.

III. Spiritual or Moral Development.

In examining this subject, we first inquire (Sect. *a*) whether there is any connection between physical and moral or religious development; then (β), what indications of moral development may be derived from history. Finally (γ), a correlation of the results of these inquiries, with the nature of the religious development in the individual, is attempted. Of course in so stupendous an inquiry but a few leading points can be presented here.

If it be true that the period of human existence on the earth has seen a gradually increasing predominance of higher motives over lower ones among the mass of mankind, and if any parts of our metaphysical being have been derived by inheritance from preëxistent beings, we are incited to the inquiry whether any of the moral qualities are included among the latter ; and whether there be any resemblance between moral and intellectual development.

Thus, if there have been a physical derivation from a preëxistent genus, and an embryonic condition of those physical characters which distinguish Homo—if there has been also an embryonic or infantile stage in intellectual qualities—we are led to inquire whether the development of the individual in moral nature will furnish us with a standard of estimation of the successive conditions or present relations of the human species in this aspect also.

a. *Relations of Physical and Moral Nature.*

Although men are much alike in the deeper qualities of their nature, there is a range of variation which is best understood by a consideration of the extremes of such variation, as seen in men of different latitudes, and women and children.

(a.) In Children. Youth is distinguished by a peculiarity, which no doubt depends upon an immature condition of the nervous center concerned, which might be called *nervous impressibility.* It is exhibited in a greater tendency to tearfulness, in timidity, less mental endurance, a greater facility in acquiring knowledge, and more ready susceptibility to the influence of sights, sounds and sensations. In both sexes the emotional nature predominates over the intelligence and judgment. In those years the *character* is said to be in embryo, and theologians in using the phrase, "reaching years of religious understanding," mean that in early years the

religious *capacities* undergo development coincidentally with those of the body.

(b.) In Women. If we examine the metaphysical characteristics of women, we observe two classes of traits—namely, those which are also found in men, and those which are absent or but weakly developed in men. Those of the first class are very similar in essential nature to those which men exhibit at an early stage of development. This may be in some way related to the fact that physical maturity occurs earlier in women.

The gentler sex is characterized by a greater impressibility, often seen in the influence exercised by a stronger character, as well as by music, color or spectacle generally; warmth of emotion, submission to its influence rather than that of logic; timidity and irregularity of action in the outer world. All these qualities belong to the male sex, as a general rule, at some period of life, though different individuals lose them at very various periods. Ruggedness and sternness may rarely be developed in infancy, yet at some still prior time they certainly do not exist in any.

Probably most men can recollect some early period of their lives when the emotional nature predominated—a time when emotion at the sight of suffering was more easily stirred than in maturer years. I do not now allude to the benevolence inspired, kept alive or developed by the influence of the Christian religion on the heart, but rather to that which belongs to the natural man. Perhaps all men can recall a period of youth when they were hero-worshipers—when they felt the need of a stronger arm, and loved to look up to the powerful friend who could sympathize with and aid them. This is the "woman stage" of character: in a large number of cases it is early passed; in some it lasts longer; while in a very few men it persists through life. Severe discipline and labor are unfavorable to its persistence. Luxury preserves its bad qualities without its good, while Christianity preserves its good elements without its bad.

It is not designed to say that woman in her emotional nature does not differ from the undeveloped man. On the contrary, though she does not differ in kind, she differs greatly in degree, for her qualities grow with her growth, and exceed in *power* many fold those exhibited by her companion at the original point of departure. Hence, since it might be said that man is the undeveloped woman, a word of explanation will be useful. Embryonic types abound in the fields of nature, but they are not therefore immature in the usual sense. Maintaining the lower essential quality, they yet exhibit the usual results of growth in individual characters; that is, increase of strength, powers of support and protection, size and beauty. In order to maintain that the masculine character coincides with that of the undeveloped woman, it would be necessary to show that the latter during her infancy possesses the male characters predominating — that is, unimpressibility, judgment, physical courage, and the like.

If we look at the second class of female characters —namely, those which are imperfectly developed or absent in men, and in respect to which man may be called undeveloped woman—we note three prominent points: facility in language, tact or finesse, and the love of children. The first two appear to me to be altogether developed results of "impressibility," already considered as an indication of immaturity. Imagination is also a quality of impressibility, and, associated with finesse, is apt to degenerate into duplicity and untruthfulness.

The third quality is different. It generally appears at a very early period of life. Who does not know how soon the little girl selects the doll, and the boy the toy-horse or machine? Here man truly never gets beyond undeveloped woman. Nevertheless, "impressibility" seems to have a great deal to do with this quality also.

Thus the metaphysical relation of the sexes would appear to be one of *inexact parallelism*, as defined in Sect. I. That the physical relation is a remote one of the same kind, several characters seem to point out. The case of the vocal organs will suffice. Their structure is identical in both sexes in early youth, and both produce nearly similar sounds. They remain in this condition in the woman, while they undergo a metamorphosis and change both in structure and vocal power in the man. In the same way, in many of the lower creation, the females possess a majority of embryonic features, though not invariably. A common example is to be found in the plumage of birds, where the females and young males are often undistinguishable.* But there are few points in the physical structure of man also in which the male condition is the

---

* Meehan states that the upper limbs and strong laterals in coniferæ and other trees produce female flowers and cones, and the lower and more interior branches the male flowers. What he points out is in harmony with the position here maintained—namely, that the female characters include more of those which are embryonic in the males, than the male characters include of those which are embryonic in the female : the female flowers are the product of the younger and more growing portions of the tree—that is, those last produced (the upper limbs and new branches)—while the male

immature one. In regard to structure, the point at which the relation between the sexes is that of *exact parallelism*, or where the mature condition of the one sex accords with the undeveloped condition of the other, is when reproduction is no longer accomplished by budding or gemmation, but requires distinct organs. Metaphysically, this relation is to be found where distinct individuality of the sexes first appears; that is, where we pass from the hermaphrodite to the bisexual condition.

But let us put the whole interpretation on this partial undevelopment of woman.

The types or conditions of organic life which have been the most prominent in the world's history—the Ganoids of the first, the Dinosaurs of the second, and the Mammoths of the third period—have generally died with their day. The line of succession has not been from them. The law of anatomy and paleontology is, that we must seek the point of departure of the type which is to predominate in the future, at lower stages on the line, in less decided forms, or in what, in scientific parlance, are called generalized types. In the same way, though the adults of the tailless apes are in a physical sense more highly developed than their young, yet the latter far more closely resemble the human species in their large facial angle and shortened jaws.

How much significance, then, is added to the law uttered by Christ!—" Except ye become as little chil-

---

flowers are produced by the older or more mature portions—that is, lower limbs or more axial regions.

Meehan's observations coincide with those of Thury and others on the origin of sexes in animals and plants, which it appears to me admit of a similar explanation, dren, ye cannot enter the kingdom of heaven." Submission of will, loving trust, confiding faith—these belong to the child: how strange they appear to the executing, commanding, reasoning man I Are they so strange to the ᐧwoman? We all know the answer. Woman is nearer to the point of departure of that development which outlives time and peoples heaven ; and if man would find it, he must retrace his steps, regain something he lost in youth, and join to the powers and energies of his character the submission, love and faith which the new birth alone can give.

Thus the summing up of the metaphysical qualities of woman would be thus expressed: In the emotional world, man's superior ; in the moral world, his equal ; in the laboring world, his inferior.

There are, however, vast differences in women in respect to the number of masculine traits they may have assumed before being determined into their own special development. Woman also, under the influence of necessity, in later years of life, may add more or less to those qualities in her which are fully developed in the man.

The relation of these facts to the principles stated as the two opposing laws of development is, it appears to me, to be explained thus : First, that woman's most inherent peculiarities are *not* the result of the external circumstances with which she has been placed in contact, as the *conflict theory* would indicate. Such circumstances are said to be her involuntary subserviency to the physically more powerful man, and the effect of a compulsory mode of life in preventing her from attaining a position of equality in the activities of the world. Second, that they *are* the result of the different distributions of qualities as already indicated by the *harmonic theory* of development; that is, of the unequal possession of features which belong to different periods in the developmental succession of the highest. And here it might be further shown that this relation involves no disadvantage to either sex, but that the principle of compensation holds in moral organization and in social order, as elsewhere. There is then another beautiful harmony which will ever remain, let the development of each sex be extended as far as it may.

(c.) In Men. If we look at the male sex, we shall find various exceptional approximations to the female in mental constitution. Further, there can be little doubt that in the Indo-European race maturity in some respects appears earlier· in tropical than in northern regions; and though subject to many exceptions, this is sufficiently general to be looked upon as a rule. Accordingly, we find in that race—at least in the warmer regions of Europe and America—a larger proportion of certain qualities which are more universal in women; as greater activity of the emotional nature when compared with the judgment; an impressibility of the nervous center, which, *cæteris paribus*, appreciates quickly the harmonies of sound, form and color; answers most quickly to the friendly greeting or the hostile menace; is more careless of consequences in the material expression of generosity or hatred, and more indifferent to truth under the influence of personal relations. The movements of the body and expressions of the countenance answer to the temperament. More of grace and elegance in the bearing mark the Greek, the Italian and the Creole, than the German, the Englishman or the Green Mountain man. More of vivacity and fire, for

better or for worse, are displayed in the countenance.

Perhaps the more northern type left all that behind in its youth. The rugged, angular character which appreciates force better than harmony, the strong intellect which delights in forethought and calculation, the less impressibility, reaching stolidity in the uneducated, are its well-known traits. If in such a character generosity is less prompt, and there is but little chivalry, there is persistency and unwavering fidelity, not readily interrupted by the lightning of passion or the dark surmises of an active imagination.

All these peculiarities appear to result, *first*, from different degrees of quickness and depth in appreciating impressions from without; and, *second*, from differing degrees of attention to the intelligent judgment in consequent action. (I leave conscience out, as not belonging to the category of inherited qualities.)

The first is the basis of an emotional nature, and the predominance of the second is the usual indication of maturity. That the first is largely dependent on an impressible condition of the nervous system can be asserted by those who reduce their nervous centers to a sensitive condition by a rapid consumption of the nutritive materials necessary to the production of thought-force, and perhaps of brain tissue itself, induced by close and prolonged mental labor. The condition of over-work, though but an imitation of immaturity, without its joy-giving nutrition, is nevertheless very instructive. The sensitiveness, both physically, emotionally and morally, is often remarkable, and a weakening of the understanding is often coincident with it.

It is necessary here to introduce a caution, that the meaning of the words high and low be not misunderstood.

Great impressibility is an essential constituent of many
of the highest forms of genius, and the combination of
this quality with strong reflective intelligence, constitutes
the most complete and efficient type of mind—therefore the highest in the common sense. It is not, however, the highest — or extremest — in an evolutional·
sense, it is not masculine, but hermaphrodite ; in other
words, its *kinetic* force exceeds its *bathmic.** It is therefore certain that a partial diminution of bathmic vigor
is an advantage to some kinds of intellect.

The above observations have been confined to the
Indo-European race. It may be objected to the theory
that savagery means immaturity in the senses above
described, as dependent largely on "impressibility,"
while savages in general display the least "impressibility," as that word is generally understood. This
cannot be asserted of the Africans, who, so far as we
know them, possess this peculiarity in a high degree.
Moreover, it must be remembered that the state of in-
difference which precedes that of impressibility in the
individual may characterize many savages ; while their
varied peculiarities may be largely accounted for by
recollecting that many combinations of different species
of emotions and kinds of intelligence go to make up
the complete result in each case.

(d.) Conclusions. Three types of religion may be
selected from the developmental conditions· of man :
first, an absence of sensibility (early infancy) ; second,
an emotional stage more productive of faith than of

---

* *Bathmic force* is analogous to the *potential* force of chemists,
but is no doubt entirely different in its nature. It is converted
into active energy or *kinetic* force only during the years of growth :
it is in large amount in *acceleration*, in small amount in *retardation*.

works; thirdly, an intellectual type, more favorable to works than to faith. Though in regard to responsibility these states may be equal, there is absolutely no gain to laboring humanity from the first type, and a serious loss in actual results from the second, taken alone, as compared with the third.

These, then, are the *physical vehicles of religion*—the *"earthen vessels"* of Paul—which give character and tone to the deeper spiritual life, as the color of the transparent vessel is communicated to the light which radiates from within.

But if evolution has taken place, there is evidently a provision for the progress from the lower to the higher states, either in the education of circumstances ("conflict,") or in the power of an interior spiritual influence ("harmony,") or both.

β. *Evidence Derived from History.*

We trace the development of Morality in—First, the family or social order; second, the civil order, or government.

Whatever may have been the extent of moral ignorance before the Deluge, it does not appear that the earth was yet prepared for the permanent habitation of the human race. All nations preserve traditions of the drowning of the early peoples by floods, such as have occurred frequently during geologic time. At the close of each period of dry land, a period of submergence has set in, and the depression of the level of the earth, and consequent overflow by the sea, has caused the death and subsequent preservation of the remains of the fauna and flora living upon it, while the elevation of

the same has produced that interruption in the process
of deposit in the same region which marks the intervals
between geologic periods. Change in these respects do
not occur to any very material extent at the present
time in the regions inhabited by the most highly developed portions of the human race ; and as the last which
occurred seems to have been expressly designed for the
preparation of the earth's surface for the occupation of
organized human society, it may be doubted whether
many such changes are to be looked for in the future.
The last great flooding was that which stratified the
drift materials of the north, and carried the finer portions far over the south, determining the minor topogra-
phy of the surface and supplying it with soils.

The existence of floods which drowned many races
of men may be considered as established. The men
destroyed by the one recorded by Moses are described
by him as exceedingly wicked, so that "the earth was
filled with violence." In his eyes the Flood was designed for their extermination.

That their condition was evil must be fully believed
if they were condemned by the executive of the Jewish
law. This law, it will be remembered, permitted polygamy, slavery, revenge, aggressive war. The Jews were
expected to rob their neighbors the Egyptians of jewels,
and they were allowed "an eye for an eye and a tooth
for a tooth." They were expected to butcher other nations, with their women and children, their flocks and
their herds. If we look at the lives of men recorded
in the Old Testament as examples of distinguished excellence, we find that their standard, however superior
to that of the people around them, would ill accord
with the morality of the present day. They were all

polygamists, slaveholders and warriors. Abraham treated Hagar and Ishmael with inhumanity. Jacob, with his mother's aid, deceived Isaac, and received thereby a blessing which extended to the whole Jewish nation. David, a man whom Paul tells us the Lord found to be after his own heart, slew the messenger who brought tidings of the death of Saul, and committed other acts which would stain the reputation of a Christian beyond redemption. It is scarcely necessary to turn to other nations if this be true of the chosen men of a chosen people. History indeed presents us with no people prior to, or contemporary with, the Jews who were not morally their inferiors.

If we turn to more modern periods, an examination of the morality of Greece and Rome reveals a curious intermixture of lower and higher moral conditions. While each of these nations produced excellent moralists, the influence of their teachings was not sufficient to elevate the masses above what would now be regarded as a very low standard. The popularity of those scenes of cruelty, the gladiatorial shows and the combats with wild beasts, sufficiently attests this. The Roman virtue of patriotism, while productive of many noble deeds, is in itself far from being a disinterested one, but partakes rather of the nature of partisanship and selfishness. If the Greeks were superior to the Romans in humanity, they were apparently their inferiors in the social virtues, and were much below the standard of Christian nations in both respects.

Ancient history points to a state of chronic war, in which the social relations were in confusion, and the development of the useful arts was almost impossible. Savage races, which continue to this day in a similar

moral condition, are, we may easily believe, most un-
happy. They are generally divided into tribes, which
are mutually hostile, or friendly only with the view of
injuring some other tribe. Might is their law, and rob-
bery, rapine and murder express their mutual relations.
This is the history of the lowest grade of barbarism,
and the history of primeval man so far as it has come
down to us in sacred and profane records. Man as a
species first appears in history as a sinful being. Then
a race maintaining a contest with the prevailing corruption and exhibiting a higher moral ideal is presented to
us in Jewish history. Finally, early Christian society
exhibits a greatly superior condition of things. In it
polygamy scarcely existed, and slavery and war were
condemned. But progress did not end here, for our
Lord said, "I have yet many things to say unto you,
but ye cannot bear 'them now. Howbeit, when He, the
spirit of truth, is come, He will guide you into all truth."

The progress revealed to us by history is truly great,
and if a similar difference existed between the first of
the human species and the first of whose condition we
have information, we can conceive how low the origin
must have been. History begins with a considerable
progress in civilization, and from this we must infer a
long preceding period of human existence, such as a
gradual evolution would require.

γ. Rationale of Moral Development.

I. *Of the Species.* Let us now look at the moral condition of the infant man of the present time. We know
his small accountability, his trust, his innocence. We
know that he is free from the law that when he "would

do good, evil is present with him," for good and evil are alike unknown. We know that until growth has progressed to a certain degree he fully deserves the praise pronounced by Our Saviour, that "of such is the kingdom of heaven." Growth, however, generally sees a change. We know that the buddings of evil appear but too soon: the lapse of a few months sees exhibitions of anger, disobedience, malice, falsehood, and their attendants—the fruit of a corruption within not manifested before.

In early youth it may be said that moral susceptibility is often in inverse ratio to physical vigor. But with growth the more physically vigorous are often sooner taught the lessons of life, for their energy brings them into earlier conflict with the antagonisms and contradictions of the world. Here is a beautiful example of the benevolent principle of compensation.

1. *Innocence and the Fall.* If physical evolution be a reality, we have reason to believe that the infantile stage of human morals, as well as of human intellect, was much prolonged in the history of our first parents. This constitutes the period of human purity, when we are told by Moses that the first pair dwelt in Eden. But the growth to maturity saw the development of all the qualities inherited from the irresponsible denizen of the forest. Man inherits from his predecessors in the creation the buddings of reason: he inherits passions, propensities and appetites. His corruption is that of his animal progenitors, and his sin is the low and bestial instinct of the brute creation. Thus only is the origin of sin made clear—a problem which the pride of man would have explained in any other way had it been possible.

But how startling the exhibition of evil by this new being as compared with the scenes of the countless ages already past! Then the right of the strongest was God's law, and rapine and destruction were the history of life. But into man had been "breathed the breath of life," and he had "become a living soul." The law of right, the Divine Spirit, was planted within him, and the laws of the beast were in antagonism to that law. The natural development of his inherited qualities necessarily brought him into collision with that higher standard planted within him, and that war was commenced which shall never cease "till He hath put all things under His feet." The first act of man's disobedience constituted the Fall, and with it would come the first *intellectual* "knowledge of good and of evil"— an apprehension up to that time derived exclusively from the divinity within, or conscience.*

2. *Free Agency.* Heretofore development had been that of physical types, but the Lord had rested on the seventh day, for man closed the line of the physical creation. Now a new development was to begin—the development of mind, of morality and of grace.

---

* In our present translation of Genesis, the Fall is ascribed to the influence of Satan assuming the form of the serpent, and this animal was cursed in consequence, and compelled to assume a prone position. This rendering may well be revised, since serpents, prone like others, existed in both America and Europe during the Eocene epoch, five times as great a period before Adam as has elapsed since his day. Clark states, with great probability, that " serpent " should be translated monkey or ape—a conclusion, it will be observed, exactly coinciding with our inductions on the basis of evolution. The instigation to evil by an ape merely states inheritance in another form. His curse, then, refers to the retention of the horizontal position by all other quadrumana, as we find it at the present day.

On the previous days of Creation all had progressed
in accordance with inevitable law apart from its objects.
Now two lines of development were at the disposal of
this being, between which his *free will* was to choose.
Did he choose the courses dictated by the spirit of the
brute, he was to be subject to the old law of the brute
creation—the right of the strongest and spiritual death.
Did he choose the guidance of the Divine Guest in his
heart, he became subject to the laws which are to guide
—I. the human species to an ultimate perfection, so far
as consistent with this world; and II. the individual
man to a higher life, where a new existence awaits him
as a spiritual being, freed from the laws of terrestrial
matter.

The charge brought against the theory of development, that it implies a necessary progress of man to all
perfection without his coöperation—or *necessitarianism*,
as it is called—is unfounded.

The free will of man remains the source alike of his
progress and his relapse.   But the choice once made,
the laws of spiritual development are apparently as inevitable as those of matter.   Thus men whose religious
capacities are increased by attention to the Divine Mon-
itor within are in the advance of progress—progress
coinciding with that which in material things is called
the *harmonic.*   On the other hand, those whose motives
are of the lower origin fall under the working of the
law of *conflict.*

The lesson derivable from the preceding considerations would seem to be "necessitarian" as respects the
whole human race, considered by itself; and I believe
it is to be truly so interpreted.   That is, the Creator of
all things has set agencies at work which will slowly

develop a perfect humanity out of His lower creation, and nothing can thwart the process or alter the result. "My word shall not return unto Me void, but it shall accomplish that which I please, and it shall prosper in the thing whereto I sent it." This is our great encouragement, our noblest hope—second only to that which looks to a blessed inheritance in another world. It is this thought that should inspire the farmer, who as he toils wonders, "Why all this labor? The Good Father could have made me like the lilies, who, though they toil not, neither spin, are yet clothed in glory; and why should I, a nobler being, be subject to the dust and the sweat of labor?" This thought should enlighten every artisan of the thousands that people the factories and guide their whirling machinery in our modern cities. Every revolution of a wheel is moving the car of progress, and the timed stroke of the crank and the rhythmic throw of the shuttle are but the music the spheres have sung since time began. A new significance then appears in the prayer of David: "Let the beauty of the Lord our God be upon us, and establish Thou the work of our hands upon us: the work of our hands, O Lord, establish Thou it." But beware of the catastrophe, for "He will sit as a refiner:" "the wheat shall be gathered into barns, but the chaff shall be burned with unquenchable fire." If this be true, let us look for—

3. *The Extinction of Evil.* How is necessitarianism to be reconciled with free will? It appears to me, thus: When a being whose safety depends on the perfection of a system of laws abandons the system by which he lives, he becomes subject to that lower grade of laws which govern lower intelligences. Man, falling from

the laws of right, comes under the dominion of the laws of brute force; as said our Saviour : " Salt is good, but if the salt have lost his savor, it is thenceforth good for nothing but to be cast forth and trodden under foot of men."

Evil, being unsatisfying to the human heart, is in its nature ever progressive, whether in the individual or the nation ; and in estimating the practical results to man of the actions prompted by the lower portion of our nature, it is only necessary to carry out to its full development each of those animal qualities which may in certain states of society be restrained by the social system. In human history those qualities have repeatedly had this development, and the battle of progress is fought to decide whether they shall overthrow the system that restrains them, or be overthrown by it.

Entire obedience to the lower instincts of our nature ensures destruction to the weaker, and generally to the stronger also. A most marked case of this kind is seen where the developed vices of civilization are introduced among a savage people—as, for example, the North American Indians. These seem in consequence to be hastening to extinction.

But a system or a circuit of existence has been allotted to the civil associations of the animal species man, independently of his moral development. It may be briefly stated thus : Races begin as poor offshoots or emigrants from a parent stock. The law of labor develops their powers, and increases their wealth and numbers. These will be diminished by their various vices ; but on the whole, in proportion as the intellectual and economical elements prevail, wealth will increase ; that is, they accumulate power, When this has been accomplished, and before activity has slackened its speed, the nation has reached the culminating point, and then it enters upon the period of decline. The restraints imposed by economy and active occupation being removed, the beastly traits find in accumulated power only increased means of gratification, and industry and prosperity sink together. Power is squandered, little is accumulated, and the nation goes down to its extinction amid scenes of internal strife and vice. Its cycle is soon fulfilled, and other nations, fresh from scenes of labor, assault it, absorb its fragments, and it dies. This has been the world's history, and it remains to be seen whether the virtues of the nations now existing will be sufficient to save them from a like fate.

Thus the history of the animal man in nations is wonderfully like that of the type or families of the animal and vegetable kingdoms during geologic ages. They rise, they increase and reach a period of multiplication and power. The force allotted to them becoming exhausted, they diminish and sink and die.

II. *Of the Individual.* In discussing physical development, we are as yet compelled to restrict ourselves to the evidence of its existence and some laws observed in the operation of its causative force. What that force is, or what are its primary laws, we know not.

So in the progress of moral development we endeavor to prove its existence and the mode of its operation, but why that mode should exist, rather than some other mode, we cannot explain.

The moral progress of the species depends, of course, on the moral progress of the individuals embraced in it. Religion is the sum of those influences which determine the motives of men's actions into harmony with the Di-

vine perfection and the Divine will. Obedience to these
influences constitutes the practice of religion, while the
statement of the growth and operation of these influences constitutes the theory of religion, or doctrine.

The Divine Spirit planted in man shows him that
which is in harmony with the Divine Mind, and it remains for his free will to conform to it or reject it. This
harmony is man's highest ideal of happiness, and in
seeking it, as well as in desiring to flee from dissonance
or pain, he but obeys the disposition common to all
conscious beings. If, however, he attempts to conform
to it, he will find the law of evil present, and frequently
obtaining the mastery. If now he be in any degree observing, he will find that the laws of morality and right
are the only ones by which human society exists in a
condition superior to that of the lower animals, and in
which the capacities of man for happiness can approach
a state of satisfaction. He may be then said to be
" awakened " to the importance of religion. If he carry
on the struggle to attain to the high goal presented to
his spiritual vision, he will be deeply grieved and hum-
bled at his failures : then he is said to be "convicted."
Under these circumstances the necessity of a deliver-
ance becomes clear, and is willingly accepted in the
only way in which it has pleased the Author of all to
present it, which has been epitomized by Paul as "the
washing of regeneration and renewal of the Holy Spirit
through Jesus Christ." Thus a life of advanced and
ever-advancing moral excellence becomes possible, and
the man makes nearer approaches to the "image of
God."

Thus is opened a new era in spiritual development,
which we are led to believe leads to an ultimate condi-

tion in which the nature inherited from our origin is entirely overcome, and an existence of moral perfection entered on. Thus in the book of Mark the simile occurs: "First the blade, then the ear, after that the full corn in the ear;" and Solomon says that the development of righteousness "shines more and more unto the perfect day." .

δ. *Summary.*

If it be true that general development in morality proceeds in spite of the original predominance of evil in the world, through the self-destructive nature of the latter, it is only necessary to examine the reasons why the excellence of the good may have been subject also to progress, and how the remainder of the race may have been influenced thereby.

The development of morality is then probably to be understood in the following sense: Since the Divine Spirit, as the prime force in moral progress, cannot in itself be supposed to have been in any way under the influence of natural laws, its capacities were no doubt as eternal and unerring in the first man as in the last. But the facts and probabilities discussed above point to development of *religious sensibility*, or capacity to appreciate moral good, or to receive impressions from the source of good.

The evidence of this is supposed to be seen in— *First*, improvement in man's views of his duty to his neighbor; and *Second*, the substitution of spiritual for symbolic religions: in other words, improvement in the capacity for receiving spiritual impressions.

What the primary cause of this supposed development of religious sensibility may have been, is a question we reverently leave untouched. That it is intimately connected in some way with, and in part dependent on, the evolution of the intelligence, appears very probable: for this evolution is seen—*First*, in a better understanding of the consequences of action, and of good and of evil in many things ; and *Second*, in the production of means for the spread of the special instrumentalities of good. The following may be enumerated as such instrumentalities :

1. Furnishing literary means of record and distribution of the truths of religion, morality and science.

2. Creating and increasing modes of transportation of teachers and literary means of disseminating truth.

3. Facilitating the migration and the spread of nations holding the highest position in the scale of morality.

4. The increase of wealth, which multiplies the extent of the preceding means.

And now, let no man attempt to set bounds to this development. Let no man say even that morality accomplished is all that is required of mankind, since that is not necessarily the evidence of a spiritual development. If a man possess the capacity for progress beyond the condition in which he finds himself, in refusing to enter upon it he declines to conform to the Divine law. And "from those to whom little is given, little is required, but from those to whom much is given, much shall be required."

SCIENTIFIC ADDRESSES.

I.

On the Methods and Tendencies of Physical Investigation.

The celebrated Fichte, in his lectures on the "Vocation of the Scholar," insisted on a culture for the scholar which should not be one-sided, but all-sided. His intellectual nature was to expand spherically, and not in a single direction. In one direction, however, Fichte required that the scholar should apply himself directly to nature, become a creator of knowledge, and thus repay, by original labors of his own, the immense debt he owed to the labors of others. It was these which enabled him to supplement the knowledge derived from his own researches, so as to render his culture rounded, and not one-sided.

Fichte's idea is to some extent illustrated by the constitution and the labors of the British Association. We have here a body of men engaged in the pursuit of natural knowledge, but variously engaged. While sympathizing with each of its departments, and supplementing his culture by knowledge drawn from all of them, each student amongst us selects one subject for the exercise of his own original faculty—one line along which he may carry the light of his private intelligence a little way into the darkness by which all knowledge is surrounded. Thus, the geologist faces the rocks ; the biologist fronts the conditions and phenomena of life ; the astronomer, stellar masses and motions ; the mathematician the properties of space and number ; the chemist pursues his atoms, while the physical investigator has his own large field in optical, thermal, electrical, acoustical, and other phenomena. The British Association, then, faces nature on all sides, and pushes knowledge centrifugally outwards, while, through circumstance or natural bent, each of its working members takes up a certain line of research in which he aspires to be an original producer, being content in all other directions to accept instruction from his fellow-men. The sum of our labors constitutes what Fichte might call the sphere of natural knowledge. In the meetings of the Association it is found necessary to resolve this sphere into its component parts, which take concrete form under the respective letters of our sections.

This section (A) is called the Mathematical and Physical section. Mathematics and Physics have been long accustomed to coalesce, and hence this grouping. For while mathematics, as a product of the human mind, is self-sustaining and nobly self-rewarding,—while the pure mathematician may never trouble his mind with considerations regarding the phenomena of the material universe, still the form of reasoning which he employs, the power which the organization of that reasoning confers, the applicability of his abstract conceptions to actual phenomena, render his science one of the most potent instruments in the solution of natural problems. Indeed, without mathematics, expressed or implied, our knowledge of physical science would be friable in the extreme.

Side by side with the mathematical method, we have the method of experiment. Here, from a starting-point furnished by his own researches or those of others, the investigator proceeds by combining intuition and verification. He ponders the knowledge he possesses and tries to push it further, he guesses and checks his guess, he conjectures and confirms or explodes his conjecture. These guesses and conjectures are by no means leaps in the dark ; for knowledge once gained casts a faint light beyond its own immediate boundaries. There is no discovery so limited as not to illuminate something beyond itself. The force of intellectual penetration into this penumbral region which surrounds actual knowledge is not dependent upon method, but is proportional to the genius of the investigator. There is, however, no genius so gifted as not to need control and verification. The profoundest minds know best that nature's ways are not at all times their ways, and that the brightest flashes in the world of thought are incomplete until they have been proved to have their counterparts in the world of fact. The vocation of the true experimentalist is the incessant correction and realization of his insight ; his experiments finally constituting a body, of which his purified intuitions are, as it were, the soul.

Partly through mathematical, and partly through experimental research, physical science has of late years assumed a momentous position in the world. Both in a material and in an intellectual point of view it has produced, and it is destined to produce, immense changes,

vast social ameliorations, and vast alterations in the
popular conception of the origin, rule, and governance
of things.   Miracles are wrought by science in the physical world, while philosophy is forsaking its ancient metaphysical channels, and pursuing those opened or indicated by scientific research.   This must become more and
more the case as philosophic writers become more deeply
imbued with the methods of science, better acquainted
with the facts which scientific men have won, and with
the great theories which they have elaborated.

If you look at the face of a watch, you see the hour
and minute-hands, and possibly also a second-hand,
moving over the graduated dial.   Why do these hands
move, and why are their relative motions such as they
are observed to be?   These questions cannot be answered without opening the watch, mastering its various
parts, and ascertaining their relationship to each other.
When this is done, we find that the observed motion of
the hands follows of necessity from the inner mechanism
of the watch when acted upon by the force invested in
the spring.

This motion of the hands may be called a phenomenon of art, but the case is similar with the phenomena
of Nature.   These also have their inner mechanism, and
their store of force to set that mechanism going.   The
ultimate problem of physical science is to reveal this
mechanism, to discern this store, and to show that from
the combined action of both, the phenomena of which
they constitute the basis must of necessity flow.

I thought that an attempt to give you even a brief and
sketchy illustration of the manner in which scientific
thinkers regard this problem would not be uninteresting
to you on the present occasion ; more especially as it

will give me occasion to say a word or two on the tendencies and limits of modern science, to point out the
region which men of science claim as their own, and
where it is mere waste of time to oppose their advance,
and also to define, if possible, the bourne between this
and that other region to which the questionings and
yearnings of the scientific intellect are directed in vain.

But here your tolerance will be needed. It was the
American Emerson, I think, who said that it is hardly
possible to state any truth strongly without apparent in-
jury to some other truth. Under the circumstances, the
proper course appears to be to state both truths strongly,
and allow each its fair share, in the formation of the resultant conviction. For truth is often of a dual character, taking the form of a magnet with two poles ; and
many of the differences which agitate the thinking part
of mankind are to be traced to the exclusiveness with
which different parties affirm one half of the duality in
forgetfulness of the other half. But this waiting for the
statement of the two sides of a question implies pa-
tience. It implies a resolution to suppress indignation if
the statement of the one half should clash with our convictions, and not to suffer ourselves to be unduly elated
if the half-statement should chime in with our views.
It implies a determination to wait calmly for the statement of the whole before we pronounce judgment either
in the form of acquiescence or dissent.

This premised, let us enter upon our task. There
have been writers who affirmed that the pyramids of
Egypt were the productions of nature ; and in his early
youth Alexander Von Humboldt wrote an essay with
the express object of refuting this notion. We now regard the pyramids as the work of men's hands, aided

probably by machinery of which no record remains. We picture to ourselves the swarming workers toiling at those vast erections, lifting the inert stones, and, guided by the volition, the skill, and possibly at times by the whip of the architect, placing the stones in their proper positions. The blocks in this case were moved by a power external to themselves, and the final form of the pyramid expressed the thought of its human builder.

Let us pass from this illustration of building power to another of a different kind. When a solution of common salt is slowly evaporated, the water which holds the salt in solution disappears, but the salt itself remains behind. At a certain stage of concentration, the salt can no longer retain the liquid form ; its particles, or molecules, as they are called, begin to deposit themselves as minute solids, so minute, indeed, as to defy all microscopic power. As evaporation continues solidification goes on, and we finally obtain, through the clustering together of innumerable molecules, a finite mass of salt of a definite form. What is this form ? It sometimes seems a mimicry of the architecture of Egypt. We have little pyramids built by the salt, terrace above terrace from base to apex, forming thus a series of steps resembling those up which the Egyptian traveler is dragged by his guides. The human mind is as little disposed to look at these pyramidal salt-crystals without further question as to look at the pyramids of Egypt without inquiring whence they came. How, then, are those salt pyramids built up ?

Guided by analogy, you may suppose that, swarming among the constituent molecules of the salt, there is an invisible population, guided and coerced by some invisible master, and placing the atomic blocks in their positions. This, however, is not the scientific idea, nor do I think your good sense will accept it as a likely one. The scientific idea is that the molecules act upon each other without the intervention of slave labor ; that they attract each other and repel each other at certain definite points, and in certain definite directions ; and that the pyramidal form is the result of this play of attraction and repulsion. While, then, the blocks of Egypt were laid down by a power external to themselves, these molecular blocks of salt are self-posited, being fixed in their places by the forces with which they act upon each other.

I take common salt as an illustration, because it is so familiar to us all ; but almost any other substance would answer my purpose equally well. In fact, throughout inorganic nature, we have this formative power, as Fichte would call it—this structural energy ready to come into play, and build the ultimate particles of matter into definite shapes. It is present everywhere. The ice of our winters and of our polar regions is its handwork, and so equally are the quartz, feldspar, and mica of our rocks. Our chalk-beds are for the most part composed of minute shells, which are also the product of structural energy ; but behind the shell, as a whole, lies the result of another and more subtle formative act. These shells are built up of little crystals of calc-spar, and to form these the structural force had to deal with the intangible molecules of carbonate of lime. This tendency on the part of matter to organize itself, to grow into shape, to assume definite forms in obedience to the definite action of force, is, as I have said, all-pervading. It is in the ground on which you tread, in the water you drink, in the air you breathe. Incipient life, in fact, manifests itself throughout the whole of what we call inorganic nature.

The forms of minerals resulting from this play of forces are various, and exhibit different degrees of complexity. Men of science avail themselves of all possible means of exploring this moleculer architecture. For this purpose they employ in turn as agents of exploration, light, heat, magnetism, electricity, and sound. Polarized light is especially useful and powerful here. A beam of such light, when sent in among the molecules of a crystal, is acted on by them, and from this action we infer with more or less of clearness the manner in which the molecules are arranged. The difference, for example, between the inner structure of a plate of rock-salt and a plate of crystalized sugar or sugar-candy is thus strikingly revealed. These differences may be made to display themselves in phenomena of color of great splendor, the play of molecular force being so regulated as to remove certain of the colored constituents of white light, and to leave others with increased intensity behind.

And now let us pass from what we are accustomed to regard as a dead mineral to a living grain of corn. When it is examined by polarized light, chromatic phenomena similar to those noticed in crystals are observed. And why? Because the architecture of the grain resembles in some degree the architecture of the crystal. In the corn the molecules are also set in definite positions, from which they act upon the light. But what has built together the molecules of the corn? I have already said, regarding crystalline architecture, that you may, if you please, consider the atoms and molecules to be placed in position by a power external to themselves.

The same hypothesis is open to you now. But, if in the case of crystals you have rejected this notion of an external architect, I think you are bound to reject it now, and to conclude that the molecules of the corn are self-posited by the forces with which they act upon each other. It would be poor philosophy to invoke an external agent in the one case and to reject it in the other.

Instead of cutting our grain into thin slices and subjecting it to the action of polarized light, let us place it in the earth and subject it to a certain degree of warmth. In other words, let the molecules, both of the corn and of the surrounding earth, be kept in a state of agitation ; for warmth, as most of you know, is, in the eye of science, tremulous molecular motion. Under these circumstances, the grain and the substances which surround it interact, and a molecular architecture is the result of this interaction. A bud is formed ; this bud reaches the surface, where it is exposed to the sun's rays, which are also to be regarded as a kind of vibratory motion. And as the common motion of heat with which the grain and the substances surrounding it were first endowed, enable the grain and these substances to coalesce, so the specific motion of the sun's rays now enables the green bud to feed upon the carbonic acid and the aqueous vapor of the air, appropriating those constituents of both for which the blade has an elective attraction, and permitting the other constituent to resume its place in the air. Thus forces are active at the root, forces are active in the blade, the matter of the earth and the matter of the atmosphere are drawn towards the plant, and the plant augments in size. We have in succession, the bud, the stalk, the ear, the full corn in the ear. For the forces here at play act in a cycle, which is completed by the production of grains similar to that with which the process began.

Now there is nothing in this process which necessarily eludes the power of mind as we know it. An intellect the same kind as our own, would, if only sufficiently expanded, be able to follow the whole process from beginning to end. No entirely new intellectual faculty would be needed for this purpose. The duly expanded mind would see in the process and its consummation an instance of the play of molecular force. It would see every molecule placed in its position by the specific attractions and repulsions exerted between it and other molecules. Nay, given the grain and its environment, an intellect the same in kind as our own, but sufficiently expanded, might trace out *à priori* every step of the process, and by the application of mechanical principles would be able to demonstrate that the cycle of actions must end, as it is seen to end, in the reproduction of forms like that with which the operation began. A similar necessity rules here to that which rules the planets in their circuits round the sun.

You will notice that I am stating my truth strongly, as at the beginning we agreed it should be stated. But I must go still further, and affirm that in the eye of science the animal body is just as much the product of molecular force as the stalk and ear of corn, or as the crystal of salt or sugar. Many of its parts are obviously mechanical. Take the human heart, for example, with its exquisite system of valves, or take the eye or the hand. Animal heat, moreover, is the same in kind as the heat of a fire, being produced by the same chemical process. Animal motion, too, is as directly derived from the food of the animal, as the motion of Treve-

thyck's walking-engine from the fuel in its furnace. As
regards matter, the animal body creates nothing; as regards force, it creates nothing. Which of you by taking thought can add one cubit to his stature? All that
has been said regarding the plant may be re-stated with
regard to the animal. Every particle that enters into
the composition of the muscle, a nerve, or a bone, has
been placed in its position by molecular force. And
unless the existence of law in these matters be denied,
and the element of caprice be introduced, we must conclude that, given the relation of any molecule of the
body to its environment, its position in the body might
be predicted. Our difficulty is not with the quality of
the problem, but with its complexity; and this difficulty
might be met by the simple expansion of the faculties
which man now possesses. Given this expansion, and
given the necessary molecular data, and the chick might
be deduced as rigorously and as logically from the egg
as the existence of Neptune was deduced from the dis-
turbances of Uranus, or as conical refraction was deduced from the undulatory theory of light.

You see I am not mincing matters, but avowing
nakedly what many scientific thinkers more or less distinctly believe. The formation of a crystal, a plant, or
an animal, is in their eyes a purely mechanical problem,
which differs from the problems of ordinary mechanics in
the smallness of the masses and the complexity of the
processes involved. Here you have one half of our
dual truth; let us now glance at the other half. Associated with this wonderful mechanism of the animal
body we have phenomena no less certain than those of
physics, but between which and the mechanism we discern no necessary connection. A man, for example,

can say I feel, I think, I love ; but how does consciousness infuse itself into the problem ? The human brain
is said to be the organ of thought and feeling ; when
we are hurt the brain feels it, when we ponder it is the
brain that thinks, when our passions or affections are
excited it is through the instrumentality of the brain.
Let us endeavor to be a little more precise here. I
hardly imagine that any profound scientific thinker who
has reflected upon the subject exists, who would not admit the extreme probability of the hypothesis, that for
every fact of consciousness, whether in the domain of
sense, of thought, or of emotion, a certain definite
molecular condition is set up in the brain ; that this relation of physics to consciousness is invariable, so that,
given the state of the brain, the corresponding thought
or feeling might be inferred ; or, given the thought or
feeling, the corresponding state of the brain might be
inferred. But how inferred ? It is at bottom not a case
of logical inference at all, but of empirical association.
You may reply that many of the inferences of science
are of this character ; the inference, for example, that
an electric current of a given direction will deflect a
magnetic needle in a definite way ; but the cases differ
in this, that the passage from the current to the needle,
if not demonstrable, is thinkable, and that we entertain
no doubt as to the final mechanical solution of the problem ; but the passage from the physics of the brain to
the corresponding facts of consciousness is unthinka-
ble. Granted that a definite thought and a definite
molecular action in the brain occur simultaneously, we
do not possess the intellectual organ, nor, apparently,
any rudiment of the organ, which would enable us to
pass by a process of reasoning from the one phenome-

non to the other.   They appear together, but we do not know why.   Were our minds and senses so expanded, strengthened, and illuminated as to enable us to see and feel the very molecules of the brain ; were we capable of following all their motions, all their groupings, all their electric discharges, if such there be ; and were we intimately acquainted with the corresponding states of thought and feeling, we should be as far as ever from the solution of the problem.   " How are these physical processes connected with the facts of consciousness ?" The chasm between the two classes of phenomena would still remain intellectually impassable.   Let the consciousness of love, for example, be associated with a right-handed spiral motion of the molecules of the brain, and the consciousness of hate with a left-handed spiral motion.   We should then know when we love that the motion is in one direction, and when we hate that the motion is in the other ; but the " WHY ?" would still remain unanswered.

In affirming that the growth of the body is mechanical, and that thought, as exercised by us, has its correlative in the physics of the brain, I think the position of the " Materialist" is stated as far as that position is a tenable one.   I think the materialist will be able finally to maintain this position against all attacks ; but I do not think, as the human mind is at present constituted, that he can pass beyond it.   I do not think he is entitled to say that his molecular groupings and his molecular motions explain everything.   In reality they explain nothing.   The utmost he can affirm is the association of two classes of phenomena of whose real bond of union he is in absolute ignorance.   The problem of the connection of the body and soul is as insoluble in its modern form as it was in the pre-scientific ages. Phosphorus is known to enter into the composition of the human brain, and a courageous writer has exclaimed, in his trenchant German, "Ohne phosphor kein gedanke." That may or may not be the case ; but even if we knew it to be the case, the knowledge would not lighten our darkness. On both sides of the zone here assigned to the materialist he is equally helpless. If you ask him whence is this "matter" of which we have been discoursing, who or what divided it into molecules, who or what impressed upon them this necessity of running into organic forms, he has no answer. Science also is mute in reply to these questions. But if the materialist is confounded, and science rendered dumb, who else is entitled to answer? To whom has the secret been revealed? Let us lower our heads and acknowledge our ignorance, one and all. Perhaps the mystery may resolve itself into knowledge at some future day. The process of things upon this earth has been one of amelioration. It is a long way from the Iguanodon and his contemporaries to the president and members of the British Association. And whether we regard the improvement from the scientific or from the theological point of view as the result of progressive development, or as the result of successive exhibitions of creative energy, neither view entitles us to assume that man's present faculties end the series—that the process of amelioration stops at him. A time may therefore come when this ultra-scientific region by which we are now enfolded may offer itself to terrestrial, if not to human investigation. Two-thirds of the rays emitted by the sun fail to arouse in the eye the sense of vision. The rays exist, but the visual organ requisite for their translation into light does not exist. And so from this region of darkness and mystery which surrounds us, rays may now be darting which require but the development of the proper intellectual organs to translate them into knowledge as far surpassing ours as ours does that of the wallowing reptiles which once held possession of this planet. Meanwhile the mystery is not without its uses. It certainly may be made a power in the human soul ; but it is a power which has feeling, not knowledge, for its base. It may be, and will be, and we hope is turned to account, both in steadying and strengthening the intellect, and in rescuing man from that littleness to which, in the struggle for existence or for precedence in the world, he is continually prone.

II.

On Haze and Dust.

Solar light in passing through a dark room reveals its track by illuminating the dust floating in the air. "The sun," says Daniel Culverwell, "discovers atomes, though they be invisible by candle-light, and makes them dance naked in his beams."

In my researches on the decomposition of vapors by light, I was compelled to remove these "atomes" and this dust. It was essential that the space containing the vapors should embrace no visible thing; that no substance capable of scattering the light in the slightest sensible degree should, at the outset of an experiment, be found in the "experimental tube" traversed by the luminous beam.

For a long time I was troubled by the appearance there of floating dust, which, though invisible in diffuse daylight, was at once revealed by a powerfully condensed beam. Two tubes were placed in succession in the path of the dust: the one containing fragments of glass wetted with concentrated sulphuric acid; the other, fragments of marble wetted with a strong solution of caustic potash. To my astonishment it passed through both. The air of the Royal Institution, sent through these tubes at a rate sufficiently slow to dry it and to remove its carbonic acid, carried into the experimental tube a considerable amount of mechanically-suspended matter, which was illuminated when the beam passed through the tube. The effect was substantially the same when the air was permitted to bubble through the liquid acid and through the solution of potash.

Thus, on the 5th of October, 1868, successive charges of air were admitted through the potash and sulphuric acid into the exhausted experimental tube. Prior to the admission of the air the tube was *optically empty;* it contained nothing competent to scatter the light. After the air had entered the tube, the conical track of the electric beam was in all cases clearly revealed. This, indeed, was a daily observation at the time to which I now refer.

I tried to intercept this floating matter in various ways ; and on the day just mentioned, prior to sending the air through the drying apparatus, I carefully permitted it to pass over the tip of a spirit-lamp flame. The floating matter no longer appeared, having been burnt up by the flame. It was, therefore, *organic matter.* When the air was sent too rapidly through the flame, a fine blue cloud was found in the experimental tube. This was the *smoke* of the organic particles. I was by no means prepared for this result ; for I had thought, with the rest of the world, that the dust of our air was, in great part, inorganic and non-combustible.

Mr. Valentin had the kindness to procure for me a small gas-furnace, containing a platinum tube, which could be heated to vivid redness. The tube also contained a roll of platinum gauze, which, while it permitted the air to pass through it, insured the practical contact of the dust with the incandescent metal. The air of the laboratory was permitted to enter the experimental tube, sometimes through the cold, and sometimes through the heated tube of platinum. · The rapidity of admission was also varied. In the first column of the following table the quantity of air operated on is expressed by the number of inches which the mercury gauge of the air-pump sank when the air entered. In the second column the condition of the platinum tube is mentioned, and in the third the state of the air which entered the experimental tube.

Quantity of Air.	State of Platinum Tube.	State of Experimental Tube.
15 inches	Cold	Full of particles.
15 "	Red-hot	Optically empty.
15 "	Cold	Full of particles.
15 "	Red-hot	Optically empty.
15 "	Cold	Full of particles.
15 "	Red-hot	Optically empty.

The phrase "optically empty" shows that when the conditions of perfect combustion were present, the floating matter totally disappeared. It was wholly burnt up, leaving not a trace of residue. From spectrum analysis, however, we know that soda floats in the air ; these organic dust particles are, I believe, the *rafts* that support it, and when they are removed it sinks and vanishes.

When the passage of the air was so rapid as to render imperfect the combustion of the floating matter, instead of optical emptiness a fine blue cloud made its appearance in the experimental tube. The following series of results illustrate this point:

Quantity.	Platinum Tube.	Experimental Tube.
15 inches, slow	Cold	Full of particles.
15 " "	Red-hot	Optically empty.
15 " quick	"	A blue cloud.
15 " "	Intensely hot	A fine blue cloud.

The optical character of these clouds was totally different from that of the dust which produced them. At right angles to the illuminating beam they discharged perfectly polarized light. The cloud could be utterly quenched by a transparent Nicol's prism, and the tube containing it reduced to optical emptiness.

The particles floating in the air of London being thus proved to be organic, I sought to burn them up at the focus of a concave reflector. One of the powerfully convergent mirrors employed in my experiments on combustion by dark rays was here made use of, but I failed in the attempt. Doubtless the floating particles are in part transparent to radiant heat, and are so far incombustible by such heat. Their rapid motion through the focus also aids their escape. They do not linger there sufficiently long to be consumed. A flame it was evident would burn them up, but I thought the presence of the flame would mask its own action among the particles.

In a cylindrical beam, which powerfully illuminated the dust of the laboratory, was placed an ignited spirit-lamp. Mingling with the flame, and round its rim, were seen wreaths of darkness resembling an intensely black smoke. On lowering the flame below the beam the same dark masses stormed upwards. They were at times blacker than the blackest smoke that I have ever seen issuing from the funnel of a steamer, and their resemblance to smoke was so perfect as to lead the most practiced observer to conclude that the apparently pure flame of the alcohol lamp required but a beam of sufficient intensity to reveal its clouds of liberated carbon.

But is the blackness smoke? The question presented itself in a moment. A red-hot poker was placed underneath the beam, and from it the black wreaths also ascended. A large hydrogen flame was next employed, and it produced those whirling masses of darkness far more copiously than either the spirit-flame or poker. Smoke was, therefore, out of the question.

What, then, was the blackness ? It was simply that of stellar space ; that is to say, blackness resulting from the absence from the track of the beam of all matter competent to scatter its light. When the flame was placed below the beam the floating matter was destroyed *in situ ;* and the air, freed from this matter, rose into the beam, jostled aside the illuminated particles and substituted for their light the darkness due to its own perfect transparency. Nothing could more forcibly illustrate the invisibility of the agent which renders all things visible. The beam crossed, unseen, the black chasm formed by the transparent air, while at both sides of the gap the thick-strewn particles shone out like a luminous solid under the powerful illumination.

But here a difficulty meets us. It is not necessary to burn the particles to produce a stream of darkness. Without actual combustion, currents may be generated which shall exclude the floating matter, and therefore appear dark amid the surrounding brightness. I noticed this effect first on placing a red-hot copper ball below the beam, and permitting it to remain there until its temperature had fallen below that of boiling water. The dark currents, though much enfeebled, were still produced. They may also be produced by a flask filled with hot water.

To study this effect a platinum wire was stretched across the beam, the two ends of the wire being connected with the two poles of a voltaic battery. To regulate the strength of the current a rheostat was placed in the circuit. Beginning with a feeble current the temperature of the wire was gradually augmented, but before it reached the heat of ignition, a flat stream of air rose from it, which when looked at edgeways appeared darker and sharper than one of the blackest lines of Fraunhofer in the solar spectrum. Right and left of this dark ·vertical band the floating matter rose upwards, bounding definitely the non-luminous stream of air. What is the explanation? Simply this. The hot wire rarefied the air in contact with it, but it did not equally lighten the floating matter. The convection current of pure air therefore passed upwards *among the particles*, dragging them after it right and left, but forming between them an impassable black partition. In this way we render an account of the dark currents produced by bodies at a temperature below that of combustion.

Oxygen, hydrogen, nitrogen, carbonic acid, so prepared as to exclude all floating particles, produce the darkness when poured or blown into the beam. Coal-gas does the same. An ordinary glass shade placed in the air with its mouth downwards permits the track of the beam to be seen crossing it. Let coal-gas or hydrogen enter the shade by a tube reaching to its top, the gas gradually fills the shade from the top downwards. As soon as it occupies the space crossed by the beam, the luminous track is instantly abolished. Lifting the shade so as to bring the common boundary of gas and air above the beam, the track flashes forth. After the shade is full, if it be inverted, the gas passes upwards like a black smoke among the illuminated particles.

The air of our London rooms is loaded with this organic dust, nor is the country air free from its pollution. However ordinary daylight may permit it to disguise itself, a sufficiently powerful beam causes the air in which the dust is suspended to appear as a semi-solid rather than as a gas. Nobody could, in the first instance, without repugnance place the mouth at the illuminated focus of the electric beam and inhale the dirt revealed there. Nor is the disgust abolished by the reflection that, although we do not see the nastiness, we are churning it in our lungs every hour and minute of our lives. There is no respite to this contact with dirt ; and the wonder is, not that we should from time to time suffer from its presence, but that so small a portion of it would appear to be deadly to man.

And what is this portion? It was some time ago the current belief that epidemic diseases generally were propagated by a kind of malaria, which consisted of organic matter in a state of *motor-decay ;* that when such matter was taken into the body through the lungs or skin, it had the power of spreading there the destroying process which had attacked itself. Such a spreading power was visibly exerted in the case of yeast. A little leaven was seen to leaven the whole lump, a mere speck of matter in this supposed state of decomposition being apparently competent to propagate indefinitely its own decay. Why should not a bit of rotten malaria work in a similar manner within the human frame? In 1836 a very wonderful reply was given to this question. In that year Cagniard de la Tour discovered the *yeast plant,* a living organism, which, when placed in a proper medium, feeds, grows, and reproduces itself, and in this way carries on the process which we name fermentation. Fermentation was thus proved to be a product of life instead of a process of decay.

Schwann, of Berlin, discovered the yeast plant independently, and in February, 1837, he also announced the important result, that when a decoction of meat is effectually screened from ordinary air, and supplied solely with air which has been raised to a high temperature, putrefaction never sets in. Putrefaction, therefore, he affirmed to be caused by something derived from the air, which something could be destroyed by a sufficiently high temperature. The experiments of Schwann were repeated and confirmed by Helmholtz and Ure. But as regards fermentation, the minds of chemists, influenced probably by the great authority of Gay-Lussac, who ascribed putrefaction to the action of oxygen, fell back upon the old notion of matter in a state of decay. It was not the living yeast plant, but the dead or dying parts of it, which, assailed by oxygen, produced the fermentation. This notion was finally exploded by Pasteur. He proved that the so-called " ferments" are not such ; that the true ferments are organized beings which find in the reputed ferments their necessary food.

Side by side with these researches and discoveries, and fortified by them and others, has run the *germ theory* of epidemic disease. The notion was expressed by Kircher, and favored by Linnæus, that epidemic diseases are due to germs which float in the atmosphere, enter the body, and produce disturbance by the development within the body of parasitic life. While it was still struggling against great odds, this theory found an expounder and a defender in the President of this Institution. At a time when most of his medical brethren considered it a wild dream, Sir Henry Holland contended that some form of the germ theory was probably true. The strength of this theory consists in the perfect parallelism of the phenomena of contagious disease with those of life. As a planted acorn gives birth to an oak competent to produce a whole crop of acorns, each gifted with the power of reproducing its parent tree, and as thus from a single seedling a whole forest may spring, so these epidemic diseases literally plant their seeds, grow, and shake abroad new germs, which, meeting in the human body their proper food and temperature, finally take possession of whole populations. Thus Asiatic cholera, beginning in a small way in the Delta of the Ganges, contrived in seventeen years to spread itself over nearly the whole habitable world. The development from an infinitesimal speck of the virus of small-pox of a crop of pustules, each charged with the original poison, is another illustration. The reappearance of the scourge, as in the case of the *Dreadnought* at Greenwich, reported on so ably by Dr. Budd and Mr. Busk, receives a satisfactory explanation from the theory which ascribes it to the lingering of germs about the infected place.

Surgeons have long known the danger of permitting air to enter an open abscess. To prevent its entrance they employ a tube called a cannula, to which is attached a sharp steel point called a trocar. They puncture with the steel point, and by gentle pressure they force the pus through the cannula. It is necessary to be very careful in cleansing the instrument ; and it is difficult to see how it can be cleansed by ordinary methods in air loaded with organic impurities, as we have proved our air to be. The instrument ought, in fact, to be made as hot as its temper will bear. But this is not done, and hence, notwithstanding all the surgeon's care, inflammation often sets in after the first operation, rendering necessary a second and a third. Rapid putrefaction is found to accompany this new inflammation. The pus, moreover, which was sweet at first, and showed no trace of animal life, is now fetid, and swarming with active little organisms called vibrios. Prof. Lister, from whose recent lecture this fact is derived, contends, with every show of reason, that this rapid putrefaction and this astounding development of animal life are due to the entry of germs into the abscess during the first operation, and their subsequent nurture and development under favorable conditions of food and temperature. The celebrated physiologist and physicist, Helmholtz, is attacked annually by hay-fever. From the 20th of May to the end of June he suffers from a catarrh of the upper air-passages ; and he has found during this period, and at no other, that his nasal secretions are peopled by these vibrios. They appear to nestle by preference in the cavities and recesses of the nose, for a strong sneeze is necessary to dislodge them.

These statements sound uncomfortable ; but by disclosing our enemy they enable us to fight him. When he clearly eyes his quarry the eagle's strength is doubled, and his swoop is rendered sure. If the germ theory be proved true, it will give a definiteness to our efforts to stamp out disease which they could not previously possess. And it is only by definite effort under its guidance that its truth or falsehood can be established. It is difficult for an outsider like myself to read without sympathetic emotion such papers as those of Dr. Budd, of Bristol, on cholera, scarlet-fever, and small-pox. He is a man of strong imagination, and may occasionally take a flight beyond his facts ; but without this dynamic heat of heart, the stolid inertia of the free-born Briton cannot be overcome. And as long as the heat is employed to warm up the truth without singeing it over-

much ; as long as this enthusiasm can overmatch its
mistakes by unequivocal examples of success, so long
am I disposed to give it a fair field to work in, and to
wish it God speed.

But let us return to our dust. It is needless to remark that it cannot be blown away by an ordinary bellows ; or, more correctly, the place of the particles
blown away is in this case supplied by others ejected
from the bellows, so that the track of the beam remains
unimpaired. But if the nozzle of a good bellows be
filled with cotton wool not too tightly packed, the air
urged through the wool is filtered of its floating matter,
and it then forms a clean band of darkness in the illuminated dust. This was the filter used by Schroëder in
his experiments on spontaneous generation, and turned
subsequently to account in the excellent researches of
Pasteur. Since 1868 I have constantly employed it
myself.

But by far the most interesting and important illustration of this filtering process is furnished by the human breath. I fill my lungs with ordinary air and
breathe through a glass tube across the electric beam.
The condensation of the aqueous vapor of the breath is
shown by the formation of a luminous white cloud of
delicate texture. It is necessary to abolish this cloud,
and this may be done by drying the breath previous to
its entering into the beam ; or still more simply, by
warming the glass tube. When this is done the luminous track of the beam is for a time uninterrupted. The
breath impresses upon the floating matter a transverse motion, but the dust from the lungs makes good the particles
displaced. But after some time an obscure disc appears
upon the beam, the darkness of which increases, until

finally, towards the end of the expiration, the beam is, as it were, pierced by an intensely black hole, in which no particles whatever can be discerned. The air, in fact, has so lodged its dirt within the lungs as to render the last portions of the expired breath absolutely free from suspended matter. This experiment may be repeated any number of times with the same result. It renders the distribution of the dirt within the lungs as manifest as if the chest were transparent.

I now empty my lungs as perfectly as possible, and placing a handful of cotton wool against my mouth and nostrils, inhale through it. There is no difficulty in thus filling the lungs with air. On expiring this air through the glass tube, its freedom from floating matter is at once manifest. From the very beginning of the act of expiration the beam is pierced by a black aperture. The first puff from the lungs abolishes the illuminated dust and puts a patch of darkness in its place, and the darkness continues throughout the entire course of the expiration. When the tube is placed below the beam and moved to and fro, the same smoke-like appearance as that obtained with a flame is observed. In short, the cotton wool, when used in sufficient quantity, completely intercepts the floating matter on its way to the lungs.

And here we have revealed to us the true philosophy of a practice followed by medical men, more from instinct than from actual knowledge. In a contagious atmosphere the physician places a handkerchief to his mouth and inhales through it. In doing so he unconciously holds back the dirt and germs of the air. If the poison were a gas it would not be thus intercepted. On showing this experiment with the cotton wool to Dr.

Bence Jones, he immediately repeated it with a silk handkerchief. The result was substantially the same, though, as might be expected, the wool is by far the surest filter. The application of these experiments is obvious. If a physician wishes to hold back from the lungs of his patient, or from his own, the germs by which contagious disease is said to be propagated, he will employ a cotton wool respirator. After the revelations of this evening, such respirators must, I think, come into general use as a defence against contagion. In the crowded dwellings of the London poor, where the isolation of the sick is difficult, if not impossible, the noxious air around the patient may, by this simple means, be restored to practical purity. Thus filtered, attendants may breathe the air unharmed. In all probability the protection of the lungs will be protection of the entire system. For it is exceedingly probable that the germs which lodge in the air-passages, and which, at their leisure, can work their way across the mucous membrane, are those which sow in the body epidemic disease. If this be so, then disease can certainly be warded off by filters of cotton wool. I should be most willing to test their efficacy in my own person. And time will decide whether in lung diseases also the woolen respirator cannot abate irritation, if not arrest decay. By its means, so far as the germs are concerned, the air of the highest Alps may be brought into the chamber of the invalid.

III.

Scientific Use of the Imagination.

I carried with me to the Alps this year the heavy burden of this evening's work. In the way of new investigation I had nothing complete enough to be brought before you ; so all that remained to me was to fall back upon such residues as I could find in the depths of consciousness, and out of them to spin the fiber and weave the web of this discourse. Save from memory I had no direct aid upon the mountains ; but to spur up the emotions, on which so much depends, as well as to nourish indirectly the intellect and will, I took with me two volumes of poetry, Goethe's " Farbenlehre," and the work on " Logic " recently published by Mr. Alexander Bain. The spur, I am sorry to say, was no match for the integument of dullness it had to pierce.

In Goethe, so glorious otherwise, I chiefly noticed the self-inflicted hurts of genius, as it broke itself in vain against the philosophy of Newton. For a time Mr. Bain became my principal companion. I found him learned and practical, shining generally with a dry light, but exhibiting at times a flush of emotional strength, which proved that even logicians share the common fire of humanity. He interested me most when he became the mirror of my own condition. Neither intellectually nor socially is it good for man to be alone, and the griefs of thought are more patiently borne when we find that they have been experienced by another. From cer-

(33)

tain passages in his book I could infer that Mr. Bain was no stranger to such sorrows. Take this passage as an illustration. Speaking of the ebb of intellectual force which we all from time to time experience, Mr Bain says: "The uncertainty where to look for the next opening of discovery brings the pain of conflict and the debility of indecision." These words have in them the true ring of personal experience.

The action of the investigator is periodic. He grapples with a subject of inquiry, wrestles with it, overcomes it, exhausts, it may be, both himself and it for the time being. He breathes a space, and then renews the struggle in another field. Now this period of halting between two investigations is not always one of pure repose. It is often a period of doubt and discomfort, of gloom and ennui. "The uncertainty where to look for the next opening of discovery brings the pain of conflict and the debility of indecision." Such was my precise condition in the Alps this year; in a score of words Mr. Bain has here sketched my mental diagnosis; and it was under these evil circumstances that I had to equip myself for the hour and the ordeal that are now come.

Gladly, however, as I should have seen this duty in other hands, I could by no means shrink from it. Disloyalty would have been worse than failure. In some fashion or other—feebly or strongly, meanly or manfully, on the higher levels of thought, or on the flats of commonplace—the task had to be accomplished. I looked in various directions for help and furtherance; but without me for a time I saw only "antres vast," and within me "deserts idle." My case resembled that of a sick doctor who had forgotten his art, and sorely needed the prescription of a friend. Mr. Bain wrote one for me. He said : " Your present knowledge must forge the links of connection between what has been already achieved and what is now required."

In these words he admonished me to review the past and recover from it the broken ends of former investigations. I tried to do so. Previous to going to Switzerland I had been thinking much of light and heat, of magnetism and electricity, of organic germs, atoms, molecules, spontaneous generation, comets and skies. With one or another of these I now sought to re-form an alliance, and finally succeeded in establishing a kind of cohesion between thought and light. The wish grew within me to trace, and to enable you to trace, some of the more occult operations of this agent. I wished, if possible, to take you behind the drop-scene of the senses, and to show you the hidden mechanism of optical action. For I take it to be well worth the while of the scientific teacher to take some pains, and even great pains, to make those whom he addresses co-partners of his thoughts. To clear his own mind in the first place from all haze and vagueness, and then to project into language which shall leave no mistake as to his meaning—which shall leave even his errors naked—the definite ideas he has shaped.

A great deal is, I think, possible to scientific exposition conducted in this way. It is possible, I believe, even before an audience like the present, to uncover to some extent the unseen things of nature, and thus to give, not only to professed students, but to others with the necessary bias, industry and capacity, an intelligent interest in the operations of science. Time and labor are necessary to this result, but science is the gainer from the public sympathy thus created.

How then are those hidden things to be revealed? How, for example, are we to lay hold of the physical basis of light, since, like that of life itself, it lies entirely without the domain of the senses? Now, philosophers may be right in affirming that we cannot transcend experience. But we can, at all events, carry it a long way from its origin. We can also magnify, diminish, qualify, and combine experiences, so as to render them fit for purposes entirely new. We are gifted with the power of imagination, combining what the Germans called *Anschauungsgabe* and *Einbildungskraft*, and by this power we can lighten the darkness which surrounds the world of the senses.

There are tories even in science who regard imagination as a faculty to be feared and avoided rather than employed. They had observed its action in weak vessels and were unduly impressed by its disasters. But they might with equal justice point to exploded boilers as an argument against the use of steam. Bounded and conditioned by coöperant reason, imagination becomes the mightiest instrument of the physical discoverer. Newton's passage from a falling apple to a falling moon was a leap of the imagination. When William Thomson tries to place the ultimate particles of matter between his compass points, and to apply to them a scale of millimeters, it is an exercise of the imagination. And in much that has been recently said about protoplasm and life, we have the outgoings of the imagination guided and controlled by the known analogies of science. In fact, without this power our knowledge of nature would be a mere tabulation of coëxistences and sequences. We should still believe in the succession of day and night, of summer and winter; but the soul of force would be dislodged from our universe; casual relations would disappear, and with them that science which is now binding the parts of nature to an organic whole.

I should like to illustrate by a few simple instances the use that scientific men have already made of this power of imagination, and to indicate afterwards some of the further uses that they are likely to make of it. Let us begin with the rudimentary experiences. Observe the falling of heavy rain drops into a tranquil pond. Each drop as it strikes the water becomes a center of . disturbance, from which a series of ring ripples expands outwards. Gravity and inertia are the agents by which this wave motion is produced, and a rough experiment will suffice to show that the rate of propagation does not amount to a foot a second.

A series of slight mechanical shocks is experienced by a body plunged in the water as the wavelets reach it in succession. But a finer motion is at the same time set up and propagated. If the head and ears be immersed in the water, as in an experiment of Franklin's, the shock of the drop is communicated to the auditory nerve—the *tick* of the drop is heard. Now this sonorous impulse is propagated, not at the rate of a foot a second, but at the rate of 4,700 feet a second. In this case it is not the gravity but the *elasticity* of the water that is the urging force. Every liquid particle pushed against its neighbor delivers up its motion with extreme rapidity, and the pulse is propagated as a thrill. The incompressibility of water, as illustrated by the famous Florentine experiment, is a measure of its elasticity, and to the possession of this property in so high a degree the rapid transmission of a sound-pulse through water is to be ascribed.

But water, as you know, is not necessary to the conduction of sound ; air is its most common vehicle. And you know that when the air possesses the particular density and elasticity corresponding to the temperature of freezing water, the velocity of sound in it is 1,090 feet a second. It is almost exactly one-fourth of the velocity in water ; the reason being that though the greater weight of the water tends to diminish the velocity, the enormous molecular elasticity of the liquid far more than atones for the disadvantage due to weight. By various contrivances we can compel the vibrations of the air to declare themselves ; we know the length and frequency of sonorous waves, and we have also obtained great mastery over the various methods by which the air is thrown into vibration. We know the phenomena and laws of vibrating rods, of organ pipes, strings, membranes, plates, and bells. We can abolish one sound by another. We know the physical meaning of music and noise, of harmony and discord. In short, as regards sound we have a very clear notion of the external physical processes which correspond to our sensations.

In these phenomena of sound we travel a very little way from downright sensible experience. Still the imagination is to some extent exercised. The bodily eye, for example, cannot see the condensations and rarefactions of the waves of sound. We construct them in thought, and we believe as firmly in their existence as in that of the air itself. But now our experience has to be carried into a new region, where a new use is to be made of it.

Having mastered the cause and mechanism of sound, we desire to know the cause and mechanism

of light. We wish to extend our inquiries from the auditory nerve to the optic nerve. Now there is in the
human intellect a power of expansion—I might almost
call it a power of creation—which is brought into play
by the simple brooding upon facts. The legend of the
Spirit brooding over chaos may have originated in a
knowledge of this power. In the case now before us it
has manifested itself by transplanting into space, for
the purposes of light, an adequately modified form of
the mechanism of sound. We know intimately whereon
the velocity of sound depends. When we lessen the
density of a medium and preserve its elasticity constant, we augment the velocity. When we highten the
elasticity and keep the density constant, we also augment the velocity. A small density, therefore, and
a great elasticity are the two things necessary to rapid
propagation.

Now light is known to move with the astounding
velocity of 185,000 miles a second. How is such a
velocity to be obtained ? By boldly diffusing in space
a medium of the requisite tenuity and elasticity. Let
us make such a medium our starting point, endowing it
with one or two other necessary qualities ; let us handle
it in accordance with strict mechanical laws ; give to
every step of your deduction the surety of the syllogism ;
carry it thus forth from the world of imagination to the
world of sense, and see whether the final outcrop of the
deduction be not the very phenomena of light which
ordinary knowledge and skilled experiment reveal. If
in all the multiplied varieties of these phenomena, including those of the most remote and entangled description, this fundamental conception always brings us face
to face with the truth ; if no contradiction to our deduc-

tions from it be found in external nature ; if, moreover, it has actually forced upon our attention phenomena which no eye had previously seen, and which no mind had previously imagined ; if by it we are gifted with a power of prescience which has never failed when brought to an experimental test ; such a conception, which never disappoints us, but always lands us on the solid shores of fact, must, we think, be something more than a mere figment of the scientific fancy. In forming it that composite and creative unity in which reason and imagination are together blent, has, we believe, led us into a world not less real than that of the senses, and of which the world of sense itself is the suggestion and justification.

Far be it from me, however, to wish to fix you immovably in this or in any other theoretic conception. With all our belief of it, it will be well to keep the theory plastic and capable of change. You may, moreover, urge that although the phenomena occur *as if* the medium existed, the absolute demonstration of its existence is still wanting. Far be it from me to deny to this reasoning such validity as it may fairly claim. Let us endeavor by means of analogy to form a fair estimate of its force.

You believe that in society you are surrounded by reasonable beings like yourself. You are, perhaps, as firmly convinced of this as of anything. What is your warrant for this conviction ? Simply and solely this, your fellow-creatures behave as if they were reasonable ; the hypothesis, for it is nothing more, accounts for the facts. To take an eminent example, you believe that our president is a reasonable being. Why ? There is no known method of superposition by which any one of us can apply himself intellectually to another so as to demonstrate coincidence as regards the possession of reason. If, therefore, you hold our president to be reasonable, it is because he behaves *as if* he were reasonable. As in the case of the ether, beyond the "*as if*" you cannot go. Nay, I should not wonder if a close comparison of the data on which both inferences rest caused many respectable persons to conclude that the ether had the best of it.

This universal medium, this light-ether as it is called, is a vehicle, not an origin of wave motion. It receives and transmits, but it does not create. Whence does it derive the motions it conveys? For the most part from luminous bodies. By this motion of a luminous body I do not mean its sensible motion, such as the flicker of a candle, or the shooting out of red prominences from the limb of the sun. I mean an intestine motion of the atoms or molecules of the luminous body. But here a certain reserve is necessary. Many chemists of the present day refuse to speak of atoms and molecules as real things. Their caution leads them to stop short of the clear, sharp, mechanically intelligible atomic theory enunciated by Dalton, or any form of that theory, and to make the doctrine of multiple proportions their intellectual bourne. I respect the caution, though I think it is here misplaced. The chemists who recoil from these notions of atoms and molecules accept without hesitation the undulatory theory of light. Like you and me they one and all believe in an ether and its light-producing waves. Let us consider what this belief involves.

Bring your imaginations once more into play and figure a series of sound waves passing through air.

Follow them up to their origin, and what do you there find? A definite, tangible, vibrating body. It may be the vocal chords of a human being, it may be an organ pipe, or it may be a stretched string. Follow in the same manner a train of ether waves to their source, remembering at the same time that your ether is matter, dense, elastic, and capable of motions subject to and determined by mechanical laws. What then do you expect to find as the source of a series of ether waves? Ask your imagination if it will accept a vibrating multiple proportion—a numerical ratio in a state of oscillation? I do not think it will. You cannot crown the edifice by this abstraction. The scientific imagination, which is here authoritative, demands as the origin and cause of a series of ether waves a particle of vibrating matter quite as definite, though it may be excessively minute, as that which gives origin to a musical sound. Such a particle we name an atom or a molecule. I think the imagination when focused so as to give definition without penumbral haze is sure to realize this image at last.

To preserve thought continuous throughout this discourse, to prevent either lack of knowledge or failure of memory from producing any rent in our picture, I here propose to run rapidly over a bit of ground which is probably familiar to most of you, but which I am anxious to make familiar to you all.

The waves generated in the ether by the swinging atoms of luminous bodies are of different lengths and amplitudes. The amplitude is the width of swing of the individual particles of the wave. In water waves it is the hight of the crest above the trough, while the length of the wave is the distance between two consecutive crests. The aggregate of waves emitted by the sun may be broadly divided into two classes, the one class competent, the other incompetent, to excite vision.

But the light-producing waves differ markedly among themselves in size, form, and force. The length of the largest of these waves is about twice that of the smallest, but the amplitude of the largest is probably a hundred times that of the smallest. Now the force or energy of the wave, which, expressed with reference to sensation, means the intensity of the light, is proportional to the square of the amplitude. Hence the amplitude being one hundred-fold, the energy of the largest light-giving waves would be ten thousand-fold that of the smallest. This is not improbable. I use these figures, not with a view to numerical accuracy, but to give you definite ideas of the differences that probably exist among the light-giving waves. And if we take the whole range of solar radiation into account—its non-visual as well as its visual waves—I think it probable that the force or energy of the largest wave is a million times that of the smallest.

Turned into their equivalents of sensation, the different light waves produce different colors. Red, for example, is produced by the largest waves, violet by the smallest, while green is produced by a wave of intermediate length and amplitude. On entering from air into more highly refracting substances, such as glass or water or the sulphide of carbon, all the waves are retarded, but the smallest ones most. This furnishes a means of separating the different classes of waves from each other—in other words, of analyzing the light. Sent through a refracting prism, the waves of the sun are turned aside in different degrees from their direct course, the red least, the violet most. They are virtually pulled asunder, and they paint upon a white screen placed to receive them "the solar spectrum."

Strictly speaking, the spectrum embraces an infinity of colors, but the limits of language and of our powers of distinction cause it to be divided into seven segments: Red, orange, yellow, green, blue, indigo, violet. These are the seven primary or prismatic colors. Separately, or mixed in various proportions, the solar waves yield all the colors observed in nature and employed in art. Collectively they give us the impression of whiteness. Pure unsifted solar light is white; and if all the wave constituents of such light be reduced in the same proportion, the light, though diminished in intensity, will still be white. The whiteness of Alpine snow with the sun shining upon it is barely tolerable to the eye. The same snow under an overcast firmament is still white. Such a firmament enfeebles the light by reflection, and when we lift ourselves above a cloud-field—to an Alpine summit, for instance, or to the top of Snowdon—and see, in the proper direction, the sun shining on the clouds, they appear dazzlingly white. Ordinary clouds, in fact, divide the solar light impinging on them into two parts—a reflected part and a transmitted part, in each of which the proportions of wave motion which produce the impression of whiteness are sensibly preserved.

It will be understood that the conditions of whiteness would fail if all the waves were diminished *equally*, or by the same absolute quantity. They must be reduced *proportionately* instead of equally. If by the act of reflection the waves of red light are split into exact halves, then, to preserve the light white, the waves of yellow, orange, green, and blue must also be split into exact halves. In short, the reduction must take place, not by absolutely equal quantities, but by equal fractional parts. In white light the preponderance as regards energy of the larger over the smaller waves must always be immense. Were the case otherwise, the physiological correlative, *blue*, of the smaller waves would have the upper hand in our sensations.

My wish to render our mental images complete, causes me to dwell briefly upon these known points, and the same wish will cause me to linger a little longer among others. But here I am disturbed by my reflections. When I consider the effect of dinner upon the nervous system, and the relation of that system to the intellectual powers I am now invoking ; when I remember that the universal experience of mankind has fixed upon certain definite elements of perfection in an after-dinner speech, and when I think how conspicuous by their absence these elements are on the present occasion, the thought is not comforting to a man who wishes to stand well with his fellow-creatures in general, and with the members of the British Association in particular. My condition might well resemble that of the ether, which is scientifically defined as an assemblage of vibrations. And the worst of it is that, unless you reverse the general verdict regarding the effect of dinner, and prove in your own persons that a uniform experience need not continue uniform—which will be a great point gained for some people—these tremors of mine are likely to become more and more painful. But I call to mind the comforting words of an inspired, though uncanonical writer, who admonishes us in the Apocrypha that fear is a bad

counsellor. Let me then cast him out, and let me trust-
fully assume that you will one and all postpone that
balmy sleep, of which dinner might, under the circumstances, be regarded as the indissoluble antecedent, and
that you will manfully and womanfully prolong your investigations of the ether and its waves into regions
which have been hitherto crossed by the pioneers of
science alone.

Not only are the waves of ether reflected by clouds,
by solids, and by liquids, but when they pass from light
air to dense, or from dense air to light, a portion of the
wave-motion is always reflected. Now our atmosphere
changes continually in density from top to bottom. It
will help our conceptions if we regard it as made up of
a series of thin concentric layers or shells of air, each
shell being of the same density throughout, and a small
and sudden change of density occuring in passing from
shell to shell. Light would be reflected at the limiting
surfaces of all these shells, and their action would be
practically the same as that of the real atmosphere.

And now I would ask your imagination to picture this
act of reflection. What must become of the reflected
light? The atmospheric layers turn their convex surfaces towards the sun ; they are so many convex mirrors of feeble power, and you will immediately perceive
that the light regularly reflected from these surfaces
cannot reach the earth at all, but is dispersed in space.

But though the sun's light is not reflected in this
fashion from the ærial layers to the earth, there is indu-
bitable evidence to show that the light of our firmament
is reflected light. Proofs of the most cogent description could be here adduced ; but we need only consider
that we receive light at the same time from all parts of

the hemisphere of heaven.   The light of the firmament
comes to us across the direction of the solar rays, and
even against the direction of the solar rays ; and this
lateral and opposing rush of wave-motion can only be
due to the rebound of the waves from the air itself, or
from something suspended in the air.   It is also evident
that, unlike the action of clouds, the solar light is not
reflected by the sky in the proportions which produce
white.   The sky is blue, which indicates a deficiency
on the part of the larger waves.   In accounting for the
color of the sky, the first question suggested by analogy
would undoubtedly be, is not the air blue ?   The blueness of the air has, in fact, been given as a solution of
the blueness of the sky.   But reason basing itself on
observation asks in reply, How, if the air be blue, can
the light of sunrise and sunset, which travels through
vast distances of air, be yellow, orange, or even red?
The passage of the white solar light through a blue medium could by no possibility redden the light.   The
hypothesis of a blue air is therefore untenable.   In fact,
the agent, whatever it is, which sends us the light of the
sky, exercises in so doing a dichroitic action.   The light
reflected is blue, the light transmitted is orange or red.
A marked distinction is thus exhibited between the matter of the sky and that of an ordinary cloud, which latter exercises no such dichroitic action.

By the force of imagination and reason combined we
may penetrate this mystery also.   The cloud takes no
note of size on the part of the waves of ether, but reflects
them all alike.   It exercises no selective action.   Now
the cause of this may be that the cloud particles are so
large in comparison with the size of the waves of ether
as to reflect them all indifferently.   A broad cliff re-

flects an Atlantic roller as easily as a ripple produced by a sea bird's wing ; and in the presence of large reflecting surfaces the existing differences of magnitude among the waves of ether may disappear. But supposing the reflecting particles, instead of being very large, to be very small, in comparison with the size of the waves. In this case, instead of the whole wave being fronted and in great part thrown back, a small portion only is shivered off. The great mass of the wave passes over such a particle without reflection. Scatter then, a handful of such minute foreign particles in our atmosphere, and set imagination to watch their action upon the solar waves. Waves of all sizes impinge upon the particles, and you see at every collision a portion of the impinging wave struck off by reflection. All the waves of the spectrum, from the extreme red to the extreme violet, are thus acted upon. But in what proportions will the waves be scattered ? A clear picture will enable us to anticipate the experimental answer. Remembering that the red waves are to the blue much in the relation of billows to ripples, let us consider whether those extremely small particles are competent to scatter all the waves in the same proportion. If they be not—and a little reflection will make it clear to you that they are not—the production of color must be an incident of the scattering. Largeness is a thing of relation ; and the smaller the wave the greater is the relative size of any particle on which the wave impinges, and the greater also the ratio of the reflected portion to the total wave.

A pebble placed in the way of the ring-ripples produced by our heavy rain-drops on a tranquil pond will throw back a large fraction of the ripple incident upon it, while the fractional part of a larger wave thrown back by the same pebble might be infinitesimal. Now we have already made it clear to our minds that to preserve the solar light white, its constituent proportions must not be altered ; but in the act of division performed by these very small particles we see that the proportions *are* altered ; an undue fraction of the smaller waves is scattered by the particles, and, as a consequence, in the scattered light blue will be the predominant color. The other colors of the spectrum must, to some extent, be associated with the blue. They are not absent, but deficient. We ought, in fact, to have them all, but in diminishing proportions, from the violet to the red.

We have here presented a case to the imagination, and assuming the undulatory theory to be a reality, we have, I think, fairly reasoned our way to the conclusion that, were particles, small in comparison to the size of the ether waves, sown in our atmosphere, the light scattered by those particles would be exactly such as we observe in our azure skies. When this light is analyzed all the colors of the spectrum are found ; but they are found in the proportions indicated by our conclusion.

Let us now turn our attention to the light which passes unscattered among the particles. How must it be finally affected ? By its successive collisions with the particles, the white light is more and more robbed of its shorter waves ; it therefore loses more and more of its due proportion of blue. The result may be anticipated. The transmitted light, where short distances are involved, will appear yellowish. But as the sun sinks towards the horizon, the atmospheric distances increase, and consequently the number of the scattering particles. They abstract, in succession, the violet, the indigo, the blue, and even disturb the proportions of green. The transmitted light under such circumstances must pass from yellow through orange to red. This also is exactly what we find in nature. Thus, while the reflected light gives us at noon the deep azure of the Alpine skies, the transmitted light gives us at sunset the warm crimson of the Alpine snows. The phenomena certainly occur *as if* our atmosphere were a medium rendered slightly turbid by the mechanical suspension of exceedingly small foreign particles.

Here, as before, we encounter our skeptical "as if." It is one of the parasites of science, ever at hand, and ready to plant itself and sprout, if it can, on the weak points of our philosophy. But a strong constitution defies the parasite, and in our case, as we question the phenomena, probability grows like growing health, until in the end the malady of doubt is completely extirpated.

The first question that naturally arises is, Can small particles be really proved to act in the manner indicated? No doubt of it. Each one of you can submit the question to an experimental test. Water will not dissolve resin, but spirit will, and when spirit which holds resin in solution is dropped into water the resin immediately separates in solid particles, which render the water milky. The coarseness of this precipitate depends on the quantity of the dissolved resin. You can cause it to separate in thick clots or in exceedingly fine particles. Professor Brücke has given us the proportions which produce particles particularly suited to our present purpose. One gramme of clean mastic is dissolved in eighty-seven grammes of absolute alcohol, and the transparent solution is allowed to drop into a beaker containing clear water kept briskly stirred. An exceedingly fine precipitate is thus formed, which declares its presence by its action upon light. Placing a dark surface behind the beaker, and permitting the light to fall into it from the top or front, the medium is seen to be distinctly blue. It is not, perhaps, so perfect a blue as I have seen on exceptional days, this year, among the Alps, but it is a very fair sky blue. A trace of soap in water gives a tint of blue. London, and I fear Liverpool milk, makes an approximation to the same color through the operation of the same cause ; and Helmholtz has irreverently disclosed the fact that a blue eye is simply a turbid medium.

Numerous instances of the kind might be cited. The action of turbid media upon light was fully and beautifully illustrated by Goethe, who, though unacquainted with the undulatory theory, was led by his experiments to regard the blue of the firmament as caused by an illuminated turbid medium with the darkness of space behind it. He describes glasses showing a bright yellow by transmitted, and a beautiful blue by reflected light. Professor Stokes, who was probably the first to discern the real nature of the action of small particles on the waves of ether, describes a glass of a similar kind. What artists call " chill " is no doubt an effect of this description. Through the action of minute particles, the browns of a picture often present the appearance of the bloom of a plum. By rubbing the varnish with a silk handkerchief optical continuity is established and the chill disappears.

Some years ago I witnessed Mr. Hirst experimenting at Zermatt on the turbid water of the Visp, which was charged with the finely divided matter ground down by the glaciers. When kept still for a day or so the grosser matter sank, but the finer matter remained suspended, and gave a distinctly blue tinge to the water. No doubt the blueness of certain Alpine lakes is in part due to this cause. Professor Roscoe has noticed several striking cases of a similar kind. In a very remarkable paper the late Principal Forbes showed that steam issuing from the safety valve of a locomotive, when favorably observed, exhibits at a certain stage of its condensation the colors of the sky. It is blue by reflected light, and orange or red by transmitted light. The effect, as pointed out by Goethe, is to some extent exhibited by peat smoke.

More than ten years ago I amused myself at Killarney, by observing on a calm day, the straight smoke columns rising from the chimneys of the cabins. It was easy to project the lower portion of a column against a bright cloud. The smoke in the former case was blue, being seen mainly by reflected light; in the latter case it was reddish, being seen mainly by transmitted light. Such smoke was not in exactly the condition to give us the glow of the Alps, but it was a step in this direction. Brücke's fine precipitate above referred to looks yellowish by transmitted light, but by duly strengthening the precipitate you may render the white light of noon as ruby colored as the sun when seen through Liverpool smoke or upon Alpine horizons.

I do not, however, point to the gross smoke arising from coal as an illustration of the action of small particles, because such smoke soon absorbs and destroys the waves of blue instead of sending them to the eyes of the observer.

These multifarious facts, and numberless others which cannot now be referred to, are explained by reference to the single principle that where the scattering particles are small in comparison to the size of the waves, we have in the reflected light a greater proportion of the smaller waves, and in the transmitted light a greater proportion of the larger waves, than existed in the original white light. The physiological consequence is that in the one light blue is predominant, and in the other light orange or red. And now let us push our inquiries forward. Our best microscopes can readily reveal objects not more than $\frac{1}{80000}$ of an inch in diameter. This is less than the length of a wave of red light. Indeed, a first-rate microscope would enable us to discern objects not exceeding in diameter the length of the smallest waves of the visible spectrum. By the microscope, therefore, we can submit our particles to an experimental test. If they are as large as the light-waves they will infallibly be seen ; and if they are not seen it is because they are smaller.

I placed in the hands of our president a bottle containing Brücke's particles in greater number and coarseness than those examined by Brücke himself. The liquid was a milky blue, and Mr. Huxley applied to it his highest microscopic power. He satisfied me at the time that had particles of even $\frac{1}{100000}$ of an inch in diameter existed in the liquid they could not have escaped detection. But no particles were seen. Under the microscope the turbid liquid was not to be distinguished from distilled water. Brücke, I may say, also found the particles to be of ultra microscopic magnitude.

But we have it in our power to imitate far more closely than we have hitherto done the natural conditions of this problem. We can generate in air, as many of you know, artificial skies, and prove their perfect identity with the natural one as regards the exhibition of a number of wholly unexpected phenomena. By a continuous process of growth, moreover, we are able to connect sky matter, if I may use the term, with molecular matter on the one side, and with molar matter, or matter in sensible masses, on the other.

In illustration of this, I will take an experiment described by M. Morren, of Marseilles, at the last meeting of the British Association. Sulphur and oxygen combine to form sulphurous acid gas. It is this choking gas that is smelt when a sulphur match is burnt in air. Two atoms of oxygen and one of sulphur constitute the molecule of sulphurous acid. Now it has been recently shown in a great number of instances that waves of ether issuing from a strong source, such as the sun or the electric light, are competent to shake asunder the atoms of gaseous molecules. A chemist would call this " decomposition" by light ; but it behooves us, who are examining the power and function of the imagination, to keep constantly before us the physical images which we hold to underlie our terms. Therefore I say, sharply and definitely, that the components of the molecules of sulphurous acid are shaken asunder by the ether waves. Inclosing the substance in a suitable vessel, placing it in a dark room, and sending through it a powerful beam of light, we at first see nothing ; the vessel containing the gas is as empty as a vacuum. Soon, however, along the track of the beam a beautiful sky-blue color is observed, which is due to the liberated particles of sulphur. For a time the blue grows more intense ; it then becomes whitish ; and from a whitish blue it passes to a more or less perfect white. If the action be continued long enough, we end by filling the tube with a dense cloud of sulphur particles, which by the application of proper means may be rendered visible.

Here, then, our ether waves untie the bond of chemical affinity, and liberate a body—sulphur—which at ordinary temperatures is a solid, and which therefore soon becomes an object of the senses. We have first of all the free atoms of sulphur, which are both invisible and incompetent to stir the retina sensibly with scattered light. But these atoms gradually coalesce and form particles, which grow larger by continual accretion until after a minute or two they appear as sky matter. In this condition they are invisible themselves, but competent to send an amount of wave motion to the retina sufficient to produce the firmamental blue. The particles continue, or may be caused to continue, in this condition for a considerable time, during which no microscope can cope with them. But they continually grow larger, and pass by insensible gradations into the state of *cloud*, when they can no longer elude the armed eye. Thus, without solution of continuity, we start with matter in the molecule, and end with matter in the mass, sky matter being the middle term of the series of transformations.

Instead of sulphurous acid we might choose from a dozen other substances, and produce the same effect with any of them. In the case of some—probably in the case of all—it is possible to preserve matter in the skyey condition for fifteen or twenty minutes under the continual operation of the light. During these fifteen or twenty minutes the particles are constantly growing larger, without ever exceeding the size requisite to the production of the celestial blue. Now when two vessels are placed before you, each containing sky matter, it is possible to state with great distinctness which vessel contains the largest particles.

The eye is very sensitive to differences of light, when, as here, the eye is in comparative darkness, and when the quantities of wave motion thrown against the retina are small. The larger particles declare themselves by the greater whiteness of their scattered light. Call now to mind the observation, or effort at observation, made by our president when he failed to distinguish the particles of resin in Brücke's medium, and when you have done so follow me. I permitted a beam of light to act upon a certain vapor. In two minutes the azure appeared, but at the end of fifteen minutes it had not ceased to be azure. After fifteen minutes, for example, its color and some other phenomena pronounced it to be a blue of distinctly smaller particles than those sought for in vain by Mr. Huxley. These particles, as already stated, must have been less than $\frac{1}{100'000}$ of an inch in diameter.

And now I want you to submit to your imagination the following question: Here are particles which have been growing continually for fifteen minutes, and at the end of that time are demonstrably smaller than those which defied the microscope of Mr. Huxley. What must have been the size of these particles at the beginning of their growth? What notion can you form of the magnitude of such particles? As the distances of stellar space give us simply a bewildering sense of vastness without leaving any distinct impression on the mind, so the magnitudes with which we have here to do impress us with a bewildering sense of smallness. We are dealing with infinitesimals compared with which the test objects of the microscope are literally immense.

From their perviousness to stellar light, and other considerations, Sir John Herschel drew some startling conclusions regarding the density and weight of comets. You know that these extraordinary and mysterious bodies sometimes throw out tails 100,000,000 of miles in length, and 50,000 miles in diameter. The diameter of our earth is 8,000 miles. Both it and the sky, and a good portion of space beyond the sky, would certainly be included in a sphere 10,000 miles across. Let us fill this sphere with cometary matter, and make it our unit of measure. An easy calculation informs us that to produce a comet's tail of the size just mentioned, about 300,000 such measures would have to be emptied into space. Now suppose the whole of this stuff to be swept together, and suitably compressed, what do you suppose its volume would be? Sir John Herschel would probably tell you that the whole mass might be carted away at a single effort by one of your dray-horses. In fact, I do not know that he would require more than a small fraction of a horse-power to remove the cometary dust. After this you will hardly regard as monstrous a notion I have sometimes entertained concerning the quantity of matter in our sky. Suppose a shell, then, to surround the earth at a hight above the surface which would place it beyond the grosser matter that hangs in the lower regions of the air—say at the hight of the Matterhorn or Mont Blanc. Outside this shell we have the deep blue firmament. Let the atmospheric space beyond the shell be swept clean, and let the sky matter be properly gathered up. What is its probable amount? I have sometimes thought that a lady's portmanteau would contain it all. I have thought that even a gentleman's portmanteau—possibly his snuff-box—might take it ın. And whether the actual sky be capable of this amount of condensation or not, I entertain no doubt that a sky quite as vast as ours, and as good in appearance, could be formed from a quantity of matter which might be held in the hollow of the hand.

Small in mass, the vastness in point of number of the particles of our sky may be inferred from the continuity of its light. It is not in broken patches nor at scattered points that the heavenly azure is revealed. To the observer on the summit of Mont Blanc the blue is as uniform and coherent as if it formed the surface of the most close-grained solid. A marble dome would not exhibit a stricter continuity. And Mr. Glaisher will inform you that if our hypothetical shell were lifted to twice the hight of Mont Blanc above the earth's surface, we should still have the azure overhead. Everywhere through the atmosphere those sky particles are strewn. They fill the Alpine valleys, spreading like a delicate gauze in front of the slopes of pine. They sometimes so swathe the peaks with light as to abolish their definition. This year I have seen the Weisshorn thus dissolved in opalescent air.

By proper instruments the glare thrown from the sky particles against the retina may be quenched, and then the mountain which it obliterated starts into sudden definition. Its extinction in front of a dark mountain resembles exactly the withdrawal of a veil. It is the light then taking possession of the eye, and not the particles acting as opaque bodies, that interfere with the definition.

By day this light quenches the stars; even by moonlight it is able to exclude from vision all stars between the fifth and the eleventh magnitude. It may be likened

to a noise, and the stellar radiance to a whisper drowned
by the noise. What is the nature of the particles which
shed this light? On points of controversy I will not
here enter, but I may say that De la Rive ascribes the
haze of the Alps in fine weather to floating organic
germs. Now the possible existence of germs in such
profusion has been held up as an absurdity. It has
been affirmed that they would darken the air, and on
the assumed impossibility of their existence in the
requisite numbers, without invasion of the solar light, a
powerful argument has been based by believers in spontaneous generation.   .

Similar arguments have been used by the opponents
of the germ theory of epidemic disease, and both parties have triumphantly challenged an appeal to the
microscope and the chemist's balance to decide the question. Without committing myself in the least to De la
Rive's notion, without offering any objection here to
the doctrine of spontaneous generation, without expressing any adherence to the germ theory of disease, I
would simply draw attention to the fact that in the atmosphere we have particles which defy both the microscope and the balance, which do not darken the air, and
which exist, nevertheless, in multitudes sufficient to reduce to insignificance the Israelitish hyperbole regarding the sands upon the seashore.

The varying judgments of men on these and other
questions may perhaps be, to some extent, accounted for
by that doctrine of relativity which plays so important
a part in philosophy. This doctrine affirms that the impressions made upon us by any circumstance, or combination of circumstances, depends upon our previous
state. Two travelers upon the same peak, the one hav-

ing ascended to it from the plain, the other having descended to it from a higher elevation, will be differently affected by the scene around them. To the one nature is expanding, to the other it is contracting, and feelings are sure to differ which have two such different antecedent states.

In our scientific judgments the law of relativity may also play an important part. To two men, one educated in the school of the senses, who has mainly occupied himself with observation, and the other educated in the school of imagination as well, and exercised in the conception of atoms and molecules to which we have so frequently referred, a bit of matter, say $\frac{1}{50000}$ of an inch in diameter, will present itself differently. The one descends to it from his molar hights, the other climbs to it from his molecular lowlands. To the one it appears small, to the other large. So also as regards the appreciation of the most minute forms of life revealed by the microscope. To one of these men they naturally appear conterminous with the ultimate particles of matter, and he readily figures the molecules from which they directly spring; with him there is but a step from the atom to the organism. The other discerns numberless organic gradations between both. Compared with his atoms, the smallest vibrios and bacteria of the microscopic field are as behemoth and leviathan.

The law of relativity may to some extent explain the different attitudes of these two men with regard to the question of spontaneous generation. An amount of evidence which satisfies the one entirely fails to satisfy the other; and while to the one the last bold defense and startling expansion of the doctrine will appear perfectly conclusive, to the other it will present itself as imposing a profitless labor of demolition on subsequent investigators. The proper and possible attitude of these two men is that each of them should work as if it were his aim and object to establish the view entertained by the other.

I trust, Mr. President, that you—whom untoward circumstances have made a biologist, but who still keep alive your sympathy with that class of inquiries which nature intended you to pursue and adorn—will excuse me to your brethren if I say that some of them seem to form an inadequate estimate of the distance which separates the microscopic from the molecular limit, and that, as a consequence, they sometimes employ a phraseology which is calculated to mislead.

When, for example, the contents of a cell are described as perfectly homogeneous, as absolutely structureless, because the microscope fails to distinguish any structure, then I think the microscope begins to play a mischievous part. A little consideration will make it plain to all of you that the microscope can have no voice in the real question of germ structure. Distilled water is more perfectly homogeneous than the contents of any possible organic germ. What causes the liquid to cease contracting at 39° F., and to grow bigger until it freezes? It is a structural process of which the microscope can take no note, nor is it likely to do so by any conceivable extension of its powers. Place this distilled water in the field of an electro-magnet, and bring a microscope to bear upon it. Will any change be observed when the magnet is excited? Absolutely none ; and still profound and complex changes have occurred.

First of all, the particles of water are rendered diamagnetically polar; and secondly, in virtue of the structure impressed upon it by the magnetic strain of its molecules, the liquid twists a ray of light in a fashion perfectly determinate both as to quantity and direction. It would be immensely interesting to both you and me if one here present, who has brought his brilliant imagination to bear upon this subject, could make us see as he sees the entangled molecular processes involved in the rotation of the plane of polarization by magnetic force. While dealing with this question he lived in a world of matter and of motion to which the microscope has no passport, and in which it can offer no aid. The cases in which similar conditions hold are simply numberless. Have the diamond, the amethyst, and the countless other crystals formed in the laboratories of nature and of man, no structure? Assuredly they have, but what can the microscope make of it? Nothing. It cannot be too distinctly borne in mind that between the microscopic limit and the true molecular limit there is room for infinite permutations and combinations. It is in this region that the poles of the atoms are arranged, that tendency is given to their powers, so that when these poles and powers have free action and proper stimulus in a suitable environment, they determine first the germ and afterwards the complete organism. This first marshaling of the atoms on which all subsequent action depends baffles a keener power than that of the microscope. Through pure excess of complexity, and long before observation can have any voice in the matter, the most highly trained intellect, the most refined and disciplined imagination, retires in bewilderment from the contemplation of the problem. We are struck dumb by an astonishment which no microscope can relieve, doubting not only the power of our instrument, but even whether we ourselves possess the intellectual elements which will ever enable us to grapple with the ultimate structural energies of nature.

But the speculative faculty, of which imagination forms so large a part, will nevertheless wander into regions where the hope of certainty would seem to be entirely shut out. We think that though the detailed analysis may be, and may ever remain, beyond us, general notions may be attainable. At all events, it is plain that beyond the present outposts of microscopic inquiry lies an immense field for the exercise of the imagination. It is only, however, the privileged spirits who know how to use their liberty without abusing it, who are able to surround imagination by the firm frontiers of reason, that are likely to work with any profit here. But freedom to them is of such paramount importance that, for the sake of securing it, a good deal of wildness on the part of weaker brethren may be overlooked. In more senses than one Mr. Darwin has drawn heavily upon the scientific tolerance of his age. He has drawn heavily upon *time* in his development of species, and he has drawn adventurously upon *matter* in his theory of pangenesis. According to this theory, a germ already microscopic is a world of minor germs. Not only is the organism as a whole wrapped up in the germ, but every organ of the organism has there its special seed.

This, I say, is an adventurous draft on the power of matter to divide itself and distribute its forces. But, unless we are perfectly sure that he is overstepping the bounds of reason, that he is unwittingly sinning against observed fact or demonstrated law—for a mind like that of Darwin can never sin wittingly against either fact or law—we ought, I think, to be cautious in limiting his intellectual horizon. If there be the least doubt in the matter, it ought to be given in favor of the freedom of such a mind. To it a vast possibility is in itself a dynamic power, though the possibility may never be drawn upon.

It gives me pleasure to think that the facts and reasonings of this discourse tend rather towards the justification of Mr. Darwin than towards his condemnation, that they tend rather to augment than to diminish the cubic space demanded by this soaring speculator; for they seem to show the perfect competence of matter and force, as regards divisibility and distribution, to bear the heaviest strain that he has hitherto imposed upon them.

In the case of Mr. Darwin, observation, imagination, and reason combined have run back with wonderful sagacity and success over a certain length of the line of biological succession. Guided by analogy, in his "Origin of Species" he placed as the root of life a primordial germ, from which he conceived the amazing richness and variety of the life that now is upon the earth's surface, might be deduced. If this were true it would not be final. The human imagination would infallibly look behind the germ, and inquire into the history of its genesis.

Certainty is here hopeless, but the materials for an opinion may be attainable. In this dim twilight of speculation the inquirer welcomes every gleam, and seeks to augment his light by indirect incidences. He studies the methods of nature in the ages and the worlds within his reach, in order to shape the course of imagination in the antecedent ages and worlds. And though the certainty possessed by experimental inquiry is here shut out, the imagination is not left entirely without guidance. From the examination of the solar system, Kant and Laplace came to the conclusion that its various bodies once formed parts of the same undislocated mass ; that matter in a nebulous form preceded matter in a dense form ; that as the ages rolled away heat was wasted, condensation followed, planets were detached, and that finally the chief portion of the fiery cloud reached, by self-compression, the magnitude and density of our sun. The earth itself offers evidence of a fiery origin ; and in our day the hypothesis of Kant and Laplace receives the independent countenance of spectrum analysis, which proves the same substances to be common to the earth and sun. Accepting some such view of the con-struction of our system as probable, a desire immediately arises to connect the present life of our planet with the past. We wish to know something of our remotest an-cestry.

On its first detachment from the central mass, life, as we understand it, could hardly have been present on the earth. How then did it come there ? The thing to be encouraged here is a reverent freedom—a freedom pre-ceded by the hard discipline which checks licentiousness in speculation—while the thing to be repressed, both in science and out of it, is dogmatism. And here I am in the hands of the meeting—willing to end, but ready to go on. I have no right to intrude upon you, unasked, the unformed notions which are floating like clouds or gathering to more solid consistency in the modern spec-ulative scientific mind. But if you wish me to speak plainly, honestly, and undisputatiously, I am willing to do so. On the present occasion

You are ordained to call, and I to come.

Two views, then, offer themselves to us. Life was present potentially in matter when in the nebulous form, and was unfolded from it by the way of natural development, or it is a principle inserted into matter at a later date. With regard to the question of time, the views of men have changed remarkably in our day and generation ; and I must say as regards courage also, and a manful willingness to engage in open contest, with fair .weapons, a great change has also occurred.

The clergy of England—at all events the clergy of London—have nerve enough to listen to the strongest views which any one amongst us would care to utter ; and they invite, if they do not challenge, men of the most decided opinions to state and stand by those opinions in open court. No theory upsets them. Let the most destructive hypothesis be stated only in the language current among gentlemen, and they look it in the face. They forego alike the thunders of heaven and the terrors of the other place, smiting the theory, if they do not like it, with honest secular strength. In fact, the greatest cowards of the present day are not to be found among the clergy, but within the pale of science itself.

Two or three years ago in an ancient London college —a clerical institution—I heard a very remarkable lecture by a very remarkable man. Three or four hundred clergymen were present at the lecture. The orator began with the civilization of Egypt in the time of Joseph ; pointing out that the very perfect organization of the kingdom, and the possession of chariots, in one of which Joseph rode, indicated a long antecedent period of civilization. He then passed on to the mud of the Nile, its rate of augmentation, its present thickness, and the remains of human handiwork found therein;

thence to the rocks which bound the Nile valley, and which team with organic remains. Thus, in his own clear and admirable way, he caused the idea of the world's age to expand itself indefinitely before the mind of his audience, and he contrasted this with the age usually assigned to the world.

During his discourse he seemed to be swimming against a stream ; he manifestly thought that he was opposing a general conviction. He expected resistance ; so did I. But it was all a mistake ; there was no adverse current, no opposing conviction, no resistance, merely here and there a half humorous but unsuccessful attempt to entangle him in his talk. The meeting agreed with all that had been said regarding the antiquity of the earth and of its life. They had, indeed, known it all long ago, and they good-humoredly rallied the lecturer for coming amongst them with so stale a story. It was quite plain that this large body of clergymen, who were, I should say, the finest samples of their class, had entirely given up the ancient landmarks, and transported the conception of life's origin to an indefinitely distant past.

In fact, clergymen, if I might be allowed a parenthesis to say so, have as strong a leaning towards scientific truth as other men, only the resistance to this bent —a resistance due to education—is generally stronger in their case than in others. They do not lack the positive element, namely, the love of truth, but the negative element, the fear of error, preponderates.

The strength of an electric current is determined by two things—the electro-motive force, and the resistance that force has to overcome. A fraction, with the former as numerator and the latter as denominator, expresses the current-strength. The "current-strength" of the clergy towards science may also be expressed by making the positive element just referred to the numerator, · and the negative one the denominator of a fraction. The numerator is not zero nor is it even small, but the denominator is large ; and hence the current strength is such as we find it to be. Slowness of conception, even open hostility, may be thus accounted for. They are for the most part errors of judgment, and not sins against truth. To most of us it may appear very simple, but to a few of us it appears transcendently wonderful, that in all classes of society truth should have this power and fascination. From the countless modifications that life has undergone through natural selection and the integration of infinitesimal steps, emerges finally the grand result that the strength of truth is greater than the strength of error, and that we have only to make the truth clear to the world to gain the world to our side. Probably no one wonders more at this result than the propounder of the law of natural selection himself. Reverting to an old acquaintance of ours, it would seem, on purely scientific grounds, as if a Veracity were at the heart of things ; as if, after ages of latent working, it had finally unfolded itself in the life of man ; as if it were still destined to unfold itself, growing in girth, throwing out stronger branches and thicker leaves, and tending more and more by its overshadowing presence to starve the weeds of error from the intellectual soil.

But this is parenthetical ; and the gist of our present inquiry regarding the introduction of life is this: Does it belong to what we call matter, or is it an independent principle inserted into matter at some suitable epoch—

say when the physical conditions become such as to permit of the development of life? Let us put the question with all the reverence due to a faith and culture in which we all were cradled—a faith and culture, moreover, which are the undeniable historic antecedents of our present enlightenment. I say, let us put the question reverently, but let us also put it clearly and definitely.

There are the strongest grounds for believing that during a certain period of its history the earth was not, nor was it fit to be, the theater of life. Whether this was ever a nebulous period, or merely a molten period, does not much matter ; and if we revert to the nebulous condition, it is because the probabilities are really on its side. Our question is this : Did creative energy pause until the nebulous matter had condensed, until the earth had been detached, until the solar fire had so far withdrawn from the earth's vicinity as to permit a crust to gather round a planet? Did it wait until the air was isolated, until the seas were formed, until evaporation, condensation, and the descent of rain had begun, until the eroding forces of the atmosphere had weathered and decomposed the molten rocks so as to form soils, until the sun's rays had become so tempered by distance and by waste as to be chemically fit for the decompositions necessary to vegetable life? Having waited through those æons until the proper conditions had set in, did it send the fiat forth, " Let life be !" ? These questions define a hypothesis not without its difficulties, but the dignity of which was demonstrated by the nobleness of the men whom it sustained.

Modern scientific thought is called upon to decide between this hypothesis and another ; and public thought generally will afterwards be called upon to do the same. You may, however, rest secure in the belief that the hypothesis just sketched can never be stormed, and that it is sure, if it yield at all, to yield to a prolonged siege. To gain new territory, modern argument requires more time than modern arms, though both of them move with greater rapidity than of yore.

But however the convictions of individuals here and there may be influenced, the process must be slow and secular which commends the rival hypothesis of natural evolution to the public mind. For what are the core and essence of this hypothesis? Strip it naked and you stand face to face with the notion that not alone the more ignoble forms of animalcular or animal life, not alone the nobler forms of the horse and lion, not alone the exquisite and wonderful mechanism of the human body, but that the human mind itself—emotion, intellect, will, and all their phenomena—were once latent in a fiery cloud. Surely the mere statement of such a motion is more than a refutation. But the hypothesis would probably go even further than this. Many who hold it would probably assent to the position that at the present moment all our philosophy, all our poetry, all our science, and all our art—Plato, Shakespeare, Newton, and Raphael—are potential in the fires of the sun.

We long to learn something of our origin. If the evolution hypothesis be correct, even this unsatisfied yearning must have come to us across the ages which separate the unconscious primeval mist from the consciousness of to-day. I do not think that any holder of the evolution hypothesis would say that I overstate it or overstrain it in any way. I merely strip it of all vagueness, and bring before you, unclothed and unvarnished, the notions by which it must stand or fall

Surely these notions represent an absurdity too monstrous to be entertained by any sane mind. Let us,
however, give them fair play. Let us steady ourselves
in front of the hypothesis, and, dismissing all terror and
excitement from our minds, let us look firmly into it with
the hard, sharp eye of intellect alone. Why are these
notions absurd, and why should sanity reject them?
The law of relativity, of which we have previously
spoken, may find its application here. These evolution
notions are absurd, monstrous, and fit only for the
intellectual gibbet in relation to the ideas concerning
matter which were drilled into us when young. Spirit
and matter have ever been presented to us in the rudest
contrast, the one as all noble, the other as all vile. But
is this correct? Does it represent what our mightiest
spiritual teacher would call the eternal fact of the universe? Upon the answer to this question all depends.

Supposing, instead of having the foregoing antithesis
of spirit and matter presented to our youthful minds, we
had been taught to regard them as equally worthy and
equally wonderful ; to consider them, in fact, as two opposite faces of the self-same mystery. Supposing that
in youth we had been impregnated with the notion of
the poet Goethe, instead of the notion of the poet
Young, looking at matter, not as brute matter, but as
"the living garment of God ;" do you not think that
under these altered circumstances the law of relativity
might have had an outcome different from its present
one? Is it not probable that our repugnance to the
idea of primeval union between spirit and matter might
be considerably abated? Without this total revolution
of the notions now prevalent the evolution hypothesis
must stand condemned ; but in many profoundly

thoughtful minds such a revolution has already taken place. They degrade neither member of the mysterious duality referred to; but they exalt one of them from its abasement, and repeal the divorce hitherto existing between both. In substance, if not in words, their position as regards spirit and matter is: "What God hath joined together let not man put asunder."

I have thus led you to the outer rim of speculative science, for beyond the nebula scientific thought has never ventured hitherto, and have tried to state that which I considered ought, in fairness, to be outspoken. I do not think this evolution hypothesis is to be flouted away contemptuously; I do not think it is to be denounced as wicked. It is to be brought before the bar of disciplined reason, and there justified or condemned. Let us hearken to those who wisely support it, and to those who wisely oppose it; and let us tolerate those, and they are many, who foolishly try to do neither of these things.

The only thing out of place in the discussion is dogmatism on either side. Fear not the evolution hypothesis. Steady yourselves in its presence upon that faith in the ultimate triumph of truth which was expressed by old Gamaliel when he said: "If it be of God, ye cannot overthrow it; if it be of man, it will come to naught." Under the fierce light of scientific inquiry this hypothesis is sure to be dissipated if it possess not a core of truth. Trust me, its existence as an hypothesis in the mind is quite compatible with the simultaneous existence of all those virtues to which the term Christian has been applied. It does not solve—it does not profess to solve—the ultimate mystery untouched. At bottom it does nothing more than "transport the conception of life's origin to an indefinitely distant past."

For, granting the nebula and its potential life, the question, whence came they? would still remain to baffle and bewilder us. And with regard to the ages of forgetfulness which lie between the conscious life of the nebula and the conscious life of the earth, it is but an extension of that forgetfulness which preceded the birth of us all. Those who hold the doctrine of evolution are by no means ignorant of the uncertainty of their data, and they yield no more to it than a provisional assent. They regard the nebular hypothesis as probable, and in the utter absence of any evidence to prove the act illegal, they extend the method of nature from the present into the past. Here the observed uniformity of nature is their only guide. Within the long range of physical inquiry they have never discerned in nature the insertion of caprice. Throughout this range the laws of physical and intellectual continuity have run side by side. Having thus determined the elements of their curve in this world of observation and experiment, they prolong that curve into an antecedent world, and accept as probable the unbroken sequence of development from the nebula to the present time.

You never hear the really philosophical defenders of the doctrine of uniformity speaking of *impossibilities* in nature. They never say, what they are constantly charged with saying, that it is impossible for the builder of the universe to alter His work. Their business is not with the possible, but the actual; not with a world which *might* be, but with a world which *is*. This they explore with a courage not unmixed with reverence, and according to methods which, like the quality of a tree, are tested by their fruits. They have but one desire— to know the truth. They have but one fear—to believe a lie. And if they know the strength of science, and rely upon it with unswerving trust, they also know the limits beyond which science ceases to be strong. They best know that questions offer themselves to thought which science, as now prosecuted, has not even the tendency to solve. They keep such questions open, and will not tolerate any unlawful limitation of the horizon of their souls. They have as little fellowship with the atheist who says there is no God as with the theist who professes to know the mind of God.

"Two things," said Immanuel Kant, "fill me with awe : the starry heavens and the sense of moral responsibility in man." And in his hours of health and strength and sanity, when the stroke of action has ceased and the pause of reflection has set in, the scientific investigator finds himself overshadowed by the same awe. Breaking contact with the hampering details of earth, it associates him with a power which gives fulness and tone to his existence, but which he can neither analyze nor comprehend.

www.ingramcontent.com/pod-product-compliance
Lightning Source LLC
Chambersburg PA
CBHW021508210326
41599CB00012B/1177